Alternative Materials for the Reinforcement and Prestressing of Concrete

Alternative Materials for the Reinforcement and Prestressing of Concrete

Edited by

JOHN L. CLARKE
Special Structures Department
Sir William Halcrow and Partners

BLACKIE ACADEMIC & PROFESSIONAL
An Imprint of Chapman & Hall
London · Glasgow · New York · Tokyo · Melbourne · Madras

Published by Blackie Academic & Professional, an imprint of Chapman & Hall, Wester Cleddens Road, Bishopbriggs, Glasgow G64 2NZ

Chapman & Hall, 2–6 Boundary Row, London SE1 8HN, UK

Blackie Academic & Professional, Wester Cleddens Road, Bishopbriggs, Glasgow G64 2NZ, UK

Chapman & Hall Inc., One Penn Plaza, 41st Floor, New York NY 10019, USA

Chapman & Hall Japan, Thomson Publishing Japan, Hirakawacho Nemoto Building, 6F, 1-7-11 Hirakawa-cho, Chiyoda-ku, Tokyo 102, Japan

DA Book (Aust.) Pty Ltd, 648 Whitehorse Road, Mitcham 3132, Victoria, Australia

Chapman & Hall India, R. Seshadri, 32 Second Main Road, CIT East, Madras 600 035, India

First edition 1993

© 1993 Chapman & Hall

Typeset in 10/12 pt Times by Best-set Typesetter Ltd., Hong Kong
Printed in Great Britain by St Edmundsbury Press, Bury St. Edmunds, Suffolk

ISBN 0 7514 0007 6

A catalogue record for this book is available from the British Library

Library of Congress Cataloging-in-Publication data
Alternative materials for the reinforcement and prestressing of
  concrete / edited by John L. Clarke.—1st ed.
      p.   cm.
      Includes bibliographical references and index.
      ISBN 0-7514-0007-6
      1. Tendons (Prestressed concrete)—Materials. 2. Reinforcing
  bars—Materials. 3. Glass fibers. 4. Polymers.   I. Clarke,
  John L., 1941–   .
  TA683.42.A57   1993
  620.1′37—dc20
                                                          93-1468
                                                              CIP

Printed on acid-free text paper, manufactured in accordance with
ANSI/NISO Z39.48-1992 (Permanence of Paper).

Preface

Concrete is probably the most widely used building material in the world. However, it does have one major limitation: while it is strong in compression, it has a very limited tensile strength. Usually, steel is used to overcome this shortcoming, either as unstressed reinforcement or in the form of prestressing tendons. If properly designed, detailed and constructed, reinforced or prestressed concrete is generally very durable. However, for structures in extremely aggressive environments, corrosion of the steel can be a significant problem. Examples of structures that may be particularly at risk include marine structures and bridges subjected to de-icing salts.

Many approaches are being tried to inhibit the corrosion mechanism in aggressive environments. Most involve protective systems of some sort, applied either to the reinforcement directly or to the exposed concrete surface. One alternative approach being developed worldwide at an increasing pace is the replacement of the steel by man-made fibres, such as carbon, glass or aramid. Generally, the fibres are encased in a resin to form a composite rod or grid. In addition, polymer grids, which have been used mostly in soils, have found applications in concrete. This book aims to bring together all aspects of the development of the alternative reinforcing materials, to describe their properties and to give details of applications worldwide, both for new structures and for the repair of existing ones. The authors have been closely associated with the development of the materials in the UK, Germany, Japan, The Netherlands, Switzerland and the USA. Between them they offer a comprehensive overview of the subject. In some areas, conflicting opinions may be expressed, but this only reflects the fact that developments are taking place on a number of different fronts at present.

The book is intended for all those concerned with concrete structures in aggressive environments—as owners, specifiers or designers—who wish to ensure adequate durability. It should be of most interest to bridge and highway authorities, and to those concerned with marine, coastal and harbour works, chemical works and water supply or drainage. Above all, it should widen the scope for designers by offering viable alternatives to conventional steel in concrete.

John L. Clarke

Contributors

Dr C.J. Burgoyne Engineering Department, Cambridge University, Trumpington Street, Cambridge CB2 21PZ, UK

Mr G.R. Carter Netlon Limited, Kelly Street, Blackburn BB2 4PJ, UK

Dr J.L. Clarke Special Structures Department, Sir William Halcrow & Partners, Vineyard House, 44 Brook Green, London W6 7BY, UK

Mr H. Deuring EMPA, Überlandstrasse 129, CH-8600 Dübendorf, Switzerland

Dr M.R. Ehsani Department of Civil Engineering, University of Arizona, Tucson, AZ 85721, USA

Mr A. Gerritse Gerritse Consultancy, Simon Huyerstraat 3, 3078 HR Rotterdam, The Netherlands

Mr H. Meier EMPA, Überlandstrasse 129, CH-8600 Dübendorf, Switzerland

Professor U. Meier EMPA, Überlandstrasse 129, CH-8600 Dübendorf, Switzerland

Mr H.-J. Miesseler SICOM GmbH, Gremberger Strasse 151a, D-51105 Köln 91, Germany

Mr G. Schwegler EMPA, Überlandstrasse 129, CH-8600 Dübendorf, Switzerland

Mr M. Sugita Shimizu Corporation, Shibaura Shimizu Building, 1-2-3 Shibaura, Minato-ku, Tokyo 105-07, Japan

Dr R. Wolff SICOM GmbH, Gremberger Strasse 151a, D-51105 Köln 91, Germany

Contents

2   Glass-fiber reinforcing bars
M.R. EHSANI

3   NEFMAC grid type reinforcement
M. SUGITA

7 Strengthening of structures with advanced composites
U. MEIER, M. DEURING, H. MEIER and
G. SCHWEGLER

153

8 Aramid-based prestressing tendons
A. GERRITSE

172

1 The need for durable reinforcement

J.L. CLARKE

1.1 Introduction

1.1.1 The mechanism of corrosion

Concrete is probably the most widely available building material in the world. It is extremely versatile and is used for all types of structures. However, it does have one major limitation: while it is strong in compression it has a very limited tensile strength. Two approaches are used to overcome this shortcoming. The first, and most common, is to use steel reinforcing bars to carry the tensile forces in the structure. The second is to prestress the concrete, using steel wires or strands. This precompresses the concrete which hence has an apparent tensile strength when subjected to bending.

If properly constructed, reinforced concrete is very durable. An adequate amount of good quality cover surrounding the reinforcement provides an alkaline environment, the steel remains in a passive state and will not corrode. However, this passivity can be destroyed either by carbonation, the steady diffusion of carbon dioxide from the atmosphere into the concrete, or by the action of chlorides. The latter may be due to contamination of the fresh concrete, for example from accelerators, or may penetrate the hardened concrete from the environment. Here the two primary sources are salt water in marine locations and de-icing salts used on highway structures.

Once the steel has become depassivated, corrosion can take place. This is a complex electro-chemical process with a flow of current from the cathodic region through the steel to the anodic region, and back through the concrete. A supply of oxygen and water, passing through the concrete, is required to fuel the cathodic reaction.

At the anode, the iron is dissolved away in the form of ferrous hydroxide. With a sufficient supply of both water and oxygen, the ferrous oxide will form rust. This occupies a volume several times that of the parent metal, the resulting expansive force causing the concrete to crack initially and then eventually spall off leaving the steel fully exposed. Figure 1.1 shows typical damage.

In general there is plenty of warning of corrosion. The first visible signs are generally cracks in the concrete, running along the line of the reinforcement, often associated with rust staining. Thus a visual survey of

Figure 1.1 Damage to a concrete post due to corrosion of the reinforcement.

the surface of the structure will give a good indication of its behaviour. Of more concern is the less common formation of so-called black rust which occurs when the supply of oxygen is limited. This is much more dense than the common red rust and does not produce the same expansive forces. Hence, there may be no visible warning of corrosion problems. This was the case in Wales in 1985 when the Ynys-y-Gwas bridge collapsed without warning [1], without even any vehicle on the structure at the time. The subsequent investigation showed that the longitudinal and transverse strands used to stress the precast units together had become severely corroded at the joints, after about 30 years in service. The inspection carried out 2 years earlier had given no indication of the problem.

1.1.2 Carbonation

The rate at which the carbonation front moves into the concrete is partly a function of the water/cement ratio of the concrete but more importantly of the amount of moisture in the concrete. For a given set of conditions, it is generally assumed that the depth of carbonation is proportional to the square root of the age of the concrete. Brown [2] carried out a visual inspection of about 100 bridges in England and Scotland ranging from 15 to more than 70 years old. From these, a representative 17 were selected for more detailed study, including assessment of carbonation depths, chloride penetration (see below) and spalling due to corrosion of the reinforcement.

He found that all the measured carbonation depths fell within the envelope of

$$D = 6\sqrt{\gamma}$$

where D is the depth in millimetres and γ is the age of the bridge in years. This would imply an overall maximum depth after 100 years of 60 mm, although 95% of the results were bounded by a curve leading to a depth after 100 years of only 40 mm. In fact, his data suggested that concretes made with modern cements carbonate more slowly than those with older cements, leading to a depth of 25 mm at 100 years for modern structures.

1.1.3 Chloride ingress

Chloride ingress is a much more rapid phenomenon than carbonation. Reviewing the results of an extensive programme of work in connection with offshore concrete oil production structures, Leeming [3] noted that chlorides penetrated rapidly down narrow cracks, of the order of 0.1 mm. (This suggests that limiting crack widths to this level in marine structures, as required by current design codes, serves no practical purpose.) Significant corrosion of the reinforcement was observed at the end of the 5-year exposure period. This rapid ingress was confirmed by Brown [2] who found significant levels of chlorides in the majority of the bridges he studied, particularly those on motorways that had been subjected to higher amounts of de-icing salts. The rate at which the corrosion occurs will depend on the availability of the oxygen and water, and is thus a function of the permeability of the concrete. In addition the environmental regime plays a major role, with alternative wetting and drying being the most severe. Thus, marine and coastal structures are particularly at risk with high levels of chloride present in the environment and a ready supply of oxygen and water.

1.1.4 The scale of the problem

The most significant cause of corrosion is chloride ingress into the concrete. This has been particularly severe for bridges where the use of de-icing salts has led to high levels of contamination. It has been estimated that in the United States, some 160 000 bridges are detrimentally affected by chloride-induced corrosion requiring a major repair programme which has been estimated at \$20 000M. In the United Kingdom, Read [4] reported on the state of the Midlands Link motorway round Birmingham.

> The 11 viaducts stretching over 21 km were originally built in 1972. Within 2 years of construction, decay was observed to have started due to winter salting of the roadway. To date (1989) some £45 million have been spent on repairs, and the total cost of construction was only £28 million when built. It is estimated that over the next 15 years, repairs will top £120 million.

Within Europe, the annual cost of corrosion has been estimated at being £1000 M per year.

1.2   Avoiding corrosion

Having identified that corrosion can be a major problem in structures in aggressive environments, what steps can be taken at the design stage to overcome the difficulty? The first step is to improve the standards of workmanship on site as many of the problems have been found to be associated with poor quality of construction. Workmanship plays a significant role in the long-term behaviour of reinforced concrete. The correct location of the reinforcement, to give an adequate cover, is essential. The concrete must be adequately compacted, to give low permeability. Exposed surfaces must be well cured, to avoid excessive drying out at early ages, which leads to the development of cracks. Excessive temperature differentials must be avoided, again to prevent the formation of cracks at an early age.

In highly aggressive environments, improvements in workmanship alone may not be sufficient to ensure adequate durability. Further steps will then be necessary. There are many possible approaches, which may be outlined as follows:

1. Improving the concrete, which may include reducing the porosity, by including admixtures to inhibit corrosion or by applying coatings to the surface to prohibit water;
2. Cathodic protection of the reinforcement, either by means of an impressed current or by sacrificial anodes;
3. Using a coated reinforcement, such as a fusion bonded epoxy coating or galvanising;
4. Using a stainless steel;
5. Using a non-ferrous reinforcement.

The aim of this chapter is to cover the provisions in codes and standards which are intended to ensure an adequate design life and then to consider the above five topics. The first four topics are looked at briefly and the fifth topic, which is covered in greater detail in the following chapters, is introduced. Where possible, some indication of the advantages and disadvantages of particular solutions are given, along with some indication of the costs that may be involved. However, no attempt has been made to recommend a best solution for any particular application. It is, of course, necessary to consider the change in the total cost of the structure, brought about by the introduction of a particular material or technique. In addition, the whole life cost, bearing in mind the cost of maintenance or repair, will give a more realistic assessment of the value of a particular solution.

Finally, although there have been limited laboratory trials on the durability of various materials, for example the work by Treadaway *et al.* [5] who compared the durability of different types of reinforcing steel, there is very little data available on the comparative in-service behaviour. What data there are will generally be qualitative rather than quantitative.

1.3 Improving the concrete

1.3.1 Requirements in codes and standards

1.3.1.1 Introduction. Codes and standards aim to achieve good durability of reinforced and prestressed structures in aggressive environments by specifying

1. high cement content;
2. low water/cement ratio;
3. suitable minimum thickness of cover to the reinforcement;
4. careful curing.

Generally, codes and standards specify a number of different exposure conditions with limiting quantities for the above, which vary depending on the severity of the environment.

In addition, many codes limit crack widths, with narrower widths for more aggressive environments. However, Beeby [6] and others have shown that, while this may influence the time to the initiation of corrosion, it will have little effect on the long-term behaviour.

1.3.1.2 British codes. In the United Kingdom, the basic standard for concrete is BS5328 [7] which covers all aspects of specification. The structural codes make reference to this document, using it as a minimum in some cases, but giving additional requirements for cover to the reinforcement.

BS5328 lists six different exposure conditions. The three most aggressive are as follows:

Very severe: surface occasionally exposed to sea water spray or de-
                 icing salts;
                 surface exposed to corrosive fumes or severe freezing
                 conditions when wet;
Most severe: surfaces frequently exposed to sea water spray or de-
                 icing salts;
Abrasive: surfaces exposed to abrasive action.

Minimum concrete grades are listed for each exposure condition; for a given grade the minimum cement content and the maximum water/cement ratio is listed; see Table 1.1.

The definitions of the various exposure conditions given in BS8110 [8], the code for structural concrete, are somewhat different, 'very severe' and 'most severe' being combined into a single category of 'very severe', while 'abrasive' is renamed 'extreme'. The minimum concrete grades are the same and are linked to requirements for nominal cover to the reinforcement. Increasing the concrete grade leads, within limits, to a reduction in the cover requirement.

There is no specific requirement for the control of crack widths in reinforced concrete members. However, for partially prestressed members, crack widths are limited to 0.2 mm in general and 0.1 mm in very aggressive environments. For fully prestressed members, no visible cracking is allowed.

Minimum periods for curing and protection are given in BS8110. The Handbook [9] points out that care should be taken in using the figures as ambient conditions may alter significantly during the period.

The maritime code, BS6349 [10] also refers to BS5328, but with more precise definitions of exposure conditions, more relevant to a marine situation. However, the minimum concrete grades are generally slightly higher, with correspondingly higher cement contents and lower water/cement ratios. The requirements for cover to the reinforcement refer back to BS8110 [8].

The bridge code, BS5400 [11] uses the same exposure conditions as in BS8110 but the definitions are once again slightly different. Crack widths are limited to 0.1 mm for extreme environments and 0.15 mm for very severe ones, for both reinforced and partially prestressed members. For fully prestressed members, no visible cracks are allowed.

Table 1.1 Mix design limits for durability in BS5328 for reinforced and prestressed concrete

Exposure condition	Maximum water/ cement ratio	Minimum cement content (kg/m^3)	Minimum grade
Very severe	0.55	325	C40
Most severe	0.45	400	C50
Abrasive	0.5	350	C45

1.3.1.3 American codes. The building code, ACI 318 [12] appears to define only one special exposure condition for reinforced concrete, covering exposure to de-icing salts, brackish water or sea water. A maximum water/cement ratio of 0.4 is given, which may be relaxed if higher than minimum covers are used. The code does not give cement contents but the Commentary gives an equivalent concrete grade of about 40. There would appear to be no specified check on crack widths.

The state-of-the-art report on concrete structures for the Arctic [13] devotes a considerable section to durability and suggests cathodic protection, epoxy-coated reinforcement, corrosion inhibitors and microsilica as possible solutions.

1.3.1.4 European code. The draft Eurocode for the design of concrete structures [14] gives five major exposure classes, some of which are further subdivided to give nine in all. The code gives only the necessary covers to the reinforcement, and limits to the permitted crack widths. For reinforced concrete, a figure of 0.3 mm is given for all but the most aggressive environment (chemical attack) for which the code says that special measures are necessary. Generally a maximum crack width of 0.2 mm is specified for prestressed concrete, with a mention of coating on the tendons for the more extreme environments. The requirements for the concrete are covered by ENV206 [15]. The latter gives minimum cement contents and maximum water/cement ratios for the various exposure classes.

1.3.1.5 Japanese code. The Japan Society of Civil Engineers specification for design [16] gives three environmental conditions. The cover increases with the severity of the exposure and with the significance of the element considered. For example, in the most severe conditions, such as marine structures, slabs require 50 mm cover, beams 60 mm and columns 70 mm.

Limiting crack widths are given which are multiples of the cover for the three different environmental conditions. The most severe requires a limiting crack width of 0.0035 times the cover. On this basis, with the covers listed above, the limiting crack widths would be 0.18 for slabs and 0.25 for columns. These are less restrictive than the British code for bridges [11]. The comments in the specification state that corrosion is not always determined by the crack width and the limiting crack width to control corrosion is not clear. This implies that the writers of the specification are aware that crack control is only one of the items required to ensure durability.

Interestingly, for special bars 'qualified as anti-corrosive or bars covered with qualified corrosion protection', the cover is determined on the basis of normal exposure, i.e. the least severe class. This would appear to be the only national document that has such a provision.

1.3.2   Reduction in permeability

1.3.2.1   Introduction.   As discussed in the previous sections, codes and standards rely on specifying a minimum cement content. The cement may be pure Portland cement or a mixture of Portland cement and ground granulated blastfurnace slag or fly ash. Tabulated values of structural properties are given for a limited range of concrete grades.

Thus the designer has a limited range in which he can work. Higher strength concretes have been developed for their improved structural properties. However, it is now realised that, because the permeability is lower, they should probably be known as high performance concretes rather than high strength concretes. (Unfortunately the latter term is likely to persist as strength is the easiest parameter to measure.)

There are two main ways of achieving reduced permeability and higher strength, namely the use of superplasticisers and the addition of microsilica to the mix.

1.3.2.2   Superplasticisers.   The requirements for a low water/cement ratio combined with the ability to place the concrete easily has led to the development of superplasticisers [17]. These make a stiff mix very workable, which is of particular benefit when the reinforcement is very congested. Alternatively the water content of the mix can be significantly reduced, leading to higher strength concretes.

Superplasticisers only act in the short term, perhaps for as little as half an hour although 3–4 h can be achieved. The mechanical properties of the hardened concrete will be very similar to those of the untreated concrete. Because of the reduced water and the improved compaction, the permeability of the superplasticised concrete will be significantly less, leading to improved protection for the reinforcement.

1.3.2.3   Microsilica.   Microsilica, otherwise known as condensed silica fume, consists of the very fine particles of amorphous silica which are collected in the dust removal systems during the manufacture of silicon and ferro-silicon metals. It has been used for a number of years in Scandinavia and North America and is now being used in a variety of applications in the United Kingdom. At first, microsilica was used mainly as a cement replacement. It is an efficient pozzolan, 1 kg replacing 3–4 kg of cement with no change in the concrete strength. Alternatively, the addition of microsilica to a standard mix will lead to significantly higher strength.

More recently, the specifiers of concrete have become more interested in the use of microsilica as an additive to give improved long-term properties. Microsilica is extremely fine, with a surface area in the region of 30 times that of normal Portland cement. The very fine particles or microspheres, pack into the small voids that usually exist in concrete. Here, they act as nucleation centres for the formation of secondary calcium silicate hydrate crystals as the calcium hydroxide in the pore solution reacts with the silicon dioxide in the microsilica. Thus, crystal growth occurs throughout the space occupied by the pore water and not just from the surface of the cement grains as in conventional concrete. The crystal growths from the microsilica spheres interact with each other and with the crystals from the cement grains. As the microspheres are highly reactive, virtually all the voids between the crystals are filled resulting in a very low permeability in the concrete.

Comparative tests have been carried out on the permeability of microsilica concrete used for various projects in the Middle East [18]. Using a range of standard German, American and British tests, microsilica concrete was shown to have low water permeability and also to have low permeability to chlorides.

Microsilica has been used extensively in the construction of the concrete oil production platforms installed in the North Sea. For example, the Gullfaks C platform built in Norway required a total of $246\,000\,\text{m}^3$ of high strength concrete (up to $80\,\text{N/mm}^2$ cube strength) containing microsilica. Figure 1.2 shows the Ninian Central Platform under construction in Scotland. Similarly the Storebelt crossing in Denmark, currently being constructed, uses microsilica at a rate of up to 8% of the total cement content. Other marine structures include wharves in Gothenburg Harbour, Sweden, built about 15 years ago and more recently coastal protection schemes in North Wales and on the East coast of England.

Microsilica has also been used for bridges, including the Tjorn Bridge in Sweden and the Rance Bridge in France. The latter achieved strengths in excess of $90\,\text{N/mm}^2$. It has also been used as an overlay on bridges in Norway where, with a strength in excess of $100\,\text{N/mm}^2$, it offers high abrasion resistance as well as protection for the reinforced concrete below.

Because the use of microsilica in concrete is relatively new, there would appear to be little information on the actual long-term in-service behaviour, but all the laboratory-based information would suggest that it should be significantly better than cconventional concrete.

1.3.3 Corrosion inhibitors

It has been shown that certain admixtures can be used to inhibit corrosion of the reinforcement in the presence of chlorides. One that shows promise is calcium nitrite.

Figure 1.2 Ninian Central oil production platform under construction.

When corrosion takes place in untreated concrete, the ferrous ions at the anode pass into solution and, in a secondary reaction, are converted to rust. With the calcium nitrite, ferric ions are formed which are insoluble and hence stay on the surface of the reinforcement, preventing further corrosion.

Berke *et al.* [19] reviewed available data from accelerated corrosion tests and concluded that:

1. The addition of calcium nitrite extends the time to corrosion initiation;
2. Total corrosion of samples with calcium nitrite is substantially less;
3. The corrosion rate, once corrosion is initiated, is less with calcium nitrite.

Although laboratory testing has been carried out since the mid-1970s, the use of calcium nitrite in actual structures appears to have been fairly small. Berke *et al.* report that about $200\,000\,m^3$ of treated concrete has been used in precast beams for bridges and parking structures, but they do not give any data on in-service behaviour.

The above, in line with most studies, have considered chloride induced corrosion; Alonso and Andrade [20] considered carbonation induced corrosion, with or without chlorides present. They concluded that, without chlorides, sodium nitrite reduced or even completely eliminated corrosion. However, with both carbonation and chlorides, the nitrites were not effective in controlling corrosion.

1.3.4 Surface treatments

In Germany and the United States, it has been common practice to seal the surface of new concrete bridges and similar structures with silanes or silane-derivatives. It has been shown from tests on control specimens that only 3% of the salt water taken up by an untreated specimen entered one treated with silane [21]. Although bridges were first treated in Germany and the United States in the early 1970s, the technique has only recently been introduced into the United Kingdom. Considerable concern has been expressed over its effectiveness with higher strength concrete [22]. While penetration into lower strength concretes is claimed to be as much as 12 mm, between 2 and 4 mm is more realistic for higher strengths. In addition, careful preparation and drying of the concrete surface is essential before application, which will be difficult in many instances. Finally, it is not clear what life can be expected from the coating and hence when it will have to be replaced.

An additional benefit suggested as a result of silane treatment is a reduced susceptibility to freeze–thaw damage because of the limited ingress of water into the concrete. Conversely, others suggest that silanes tend to trap the existing water in the concrete, which cannot escape when it starts to freeze, leading to more damage than would occur with untreated concrete. However, Perenchio [21] surveyed 37 bridges treated with silanes some 15 years earlier, and concluded that there was no significant effect, silane treatment neither increasing nor decreasing the amount of freeze–thaw damage.

Although not strictly a surface treatment, permeable formwork is used in some situations. This lowers the water/cement ratio at the surface leading to a much reduced permeability of the concrete in the region.

1.3.5 Grouting of prestressing tendons

In post-tensioned structures, the prestressing tendons or wires pass through ducts or preformed voids in the concrete. After stressing, the ducts are filled, usually with a cementitious grout. The intention is to provide an alkali protection to the steel as well as ensuring composite behaviour with the concrete.

In practice, it has been found that insufficient care is sometimes taken with the grouting process, resulting in sections of the duct being incompletely filled. This not only leads to little or no protection for the steel but can also form pockets in which water can accumulate. In a bridge deck, this water is likely to be heavily contaminated with chlorides from de-icing salts, increasing the likelihood of corrosion. This situation led to the collapse of the Ynys-y-Gwas bridge in Wales [1] where insufficient attention had been paid to the grouting of the ducts at the joints between the precast units. In correctly grouted regions, the tendons were un-corroded after 30 years in service while at the joints, they had corroded so severely that they failed under the self-weight of the bridge.

There are a number of ways in which the grout may be improved, including using expansive agents to encourage complete filling of the duct and the inclusion of corrosion inhibitors. However, these are unlikely to be adequate substitutes for good workmanship coupled with adequate supervision.

1.4   Cathodic protection

1.4.1   Introduction

As discussed in Section 1.1.1, the corrosion of steel reinforcement in concrete is an electro-chemical process. Cathodic protection is a technique by which the electrical potential of the steel is increased to a level at which corrosion cannot take place. It is widely used for both steel and concrete offshore structures, while on land it has been used for the protection of pipelines and similar structures. It has been used, on a limited scale, for concrete structures as discussed below.

Two different methods are employed, an impressed current and the use of sacrificial anodes. In the first, the structure is connected to the negative terminal of a DC power source, ideally using an anode which does not corrode. In the second, the reinforcement is connected to anodes with a more negative corrosion potential than steel, such as zinc or aluminium. The current is reversed and corrosion now takes place at the anode, which is gradually used up. In both cases, electrical continuity of the reinforcement is required.

Most concrete applications to date have been for the repair and rehabilitation of structures that have already suffered corrosion damage. A number of different techniques have been developed for providing a suitable anode; these are outlined in the following section. In addition, the principles of cathodic protection have been used to remove chlorides from contaminated concrete.

1.4.2   Provision of anodes

It has been found in practice that the anode must be fitted to the concrete covering most of the surface area. This ensures uniform protection of the embedded steel. There are a number of different methods of forming an anode; they are presented here in no particular order. The first is copper wire, in a polymer sheath, which is fixed to the surface of the concrete in a serpentine arrangement. Alternatively, a coated titanium mesh may be

Figure 1.3 Installation of cathodic protection anodes.

used, which gives a more uniform covering to the surface. Simple clips are used to hold the anode in place. Figure 1.3 shows a system applied to a column and the soffit of a beam. After installation, the whole area is covered with an overlay. This may be a cementitious grout or concrete depending on the particular application. For vertical surfaces, sprayed concrete is often used because of its ease of application.

In general, the application of cathodic protection will be part of rehabilitation work and hence it is likely that chloride contaminated or carbonated concrete will have been removed. Thus, the overlay serves the additional function of reinstating the cover to the reinforcement. Alternatively, if the surface concrete does not have to be removed, the anode wires can be set in grooves cut into the concrete which are then filled with a suitable grout.

Where surfaces will not be subjected to abrasion, thin film anodes can be applied directly to the concrete. These include conductive paints and spray applied metals. The paints use graphite as the electrically conductive agent in a variety of bases to form the film. Although they are simple to apply, the paints have a limited capacity to conduct large currents and are easily damaged requiring regular maintenance.

A similar approach is to use a sprayed zinc coating. This acts as a sacrificial anode which is gradually consumed by the anodic reaction. Its rate of degradation can be monitored and a further layer added as required. The electric current is provided by individual wires which are either titanium or stainless steel.

The final method is to provide discrete anodes. These consist of suitable plates bonded to the surface on a regular grid and each fed by an individual power cable. This approach is only suitable for fairly limited areas of application.

1.4.3   Applications

Cathodic protection has been applied to a number of bridges in the United States and Canada [23] using a range of different techniques. Recently large scale comparative trials were carried out in the United Kingdom on the Midlands Link elevated motorway near Birmingham. These led to the conclusion that, for this particular application, conductive paint systems were more suitable than ones using mesh [24]. It is understood that 80 support structures are scheduled for cathodic protection and in the region of $10\,000\,m^2$ of concrete surface have already been applied.

In Bahrain, the Manama-Sitra Causeway, built in 1975, showed such severe corrosion after about 10 years that the deck became structurally unsafe and had to be replaced. The pile caps were repaired and cathodically protected with a mesh and overlay system.

While the majority of applications of cathodic protection have been for bridges, other types of structure have been protected. Examples include a 14-storey residential tower block in Cardiff where plasticised titanium wires were set into grooves cut in the surface of the walls.

The first installation in the United Kingdom of cathodic protection for a new concrete structure was for the concrete deck on a roll-on/roll-off ramp at Felixstowe Dock. The system was commissioned in January 1992.

1.4.4   Costs

Weyers and Cady [23] give costs based on North American experience of cathodic protection. They suggested a cost over a 40-year period of $165/m^2$ (at 1984 prices) for a wire-anode system without an overlay. On this basis, they suggested that cathodic protection was not yet economically competitive with conventional repair techniques. However, as a result of the Midlands Link trials and further studies, it has been concluded that cathodic protection would save £140M over a 10-year period [24]. This is on the basis of installed costs of about £100/m². The chief difficulty in determining the true cost of cathodic protection systems is in assessing the life of the anode, which is the most expensive part of the system. Many of those being used at present have been developed within the last 10 years and hence it is difficult to determine their long-term behaviour.

1.4.5   Removing chlorides from the concrete

A process similar to cathodic protection can be used to remove chlorides from concrete. A temporary anode system, consisting of an absorbent layer with conducting mat, is laid on the surface and a current applied at a much higher level than would be used normally for cathodic protection. The chlorides are gradually transferred to the absorbent layer and can hence be removed. To date little reliable information on the effectiveness of the system is available, but it would appear to have great potential, particularly for areas where the cost of repair is high.

1.5   Fusion-bonded epoxy-coated reinforcement

1.5.1   Introduction

In the early 1970s, many North American bridges were starting to show signs of corrosion soon after construction and were frequently requiring extensive repair after only 10 years or so. This was largely caused by the use of de-icing salts applied to the roadway, made worse by the fact that bridges at that time were constructed without a waterproof layer below the running surface. The Federal Highways Administration carried out a study of a range of coatings [25]. The aim was to find one that would form a sufficiently durable barrier to aggressive materials such as chlorides and yet be robust enough to tolerate handling, fixing and the pouring of concrete. It was found that fusion-bonded epoxy coatings were the most suitable. The first highway bridge using fusion-bonded epoxy-coated reinforcement was built in Pennsylvania in 1973. Since that time, its use has spread rapidly, with coating plants being set up in many countries. Subsequent sections give details of the manufacturing process, its use in practice, actual applications and possible limitations.

1.5.2   Production

In the United Kingdom, the coating of the reinforcement is covered by BS7295 [26] and in the United States by ASTM A775 [27]. Plants are designed to coat straight bars in a continuous process. Initially, the bar is shot blasted to remove any mill-scale and to give the surface finish required by the appropriate Standard. This ensures an adequate bond between the epoxy and the steel. The bar is then heated to a carefully controlled temperature before passing through a spray booth. Here, electrostatically charged epoxy powder particles are deposited evenly on the surface of the bar. The process gives a uniform powder coating that flows over the surface of the bar. The thickness is closely controlled to ensure that the bar is adequately protected and yet the surface characteristics, which will influence its bond capacity once embedded in concrete, are not unduly affected.

Shortly after spraying, the epoxy coating starts to cure and hardens sufficiently for the bar to be handled. The resin continues to cure on the bar and cools to ambient temperature. On completion, the coating is tested, specifically checking that the thickness is within specified limits (typically 130–300 μm) and that there are few holidays or gaps in the coating (the British Standard allows a maximum of 5/m).

1.5.3   Cutting, bending and installation

The coating is tough enough to resist normal impact during handling and the standards include requirements for bending the bars to a given radius without damage. To minimise the risk of damage, British practice is to supply coated reinforcement already cut and bent. This allows the supplier to use padded contact areas on his bending machine and a controlled bending rate, all carried out at a suitable temperature, to minimise damage. Cut ends are protected with an epoxy repair compound and any accidental damage to the coating is similarly covered. Bars are tied together into bundles, with protective padding to avoid damage to the coating.

Reasonable care is required with handling and storing the bars on site. There are two causes of damage to be avoided. The first is mechanical damage to the coating, for example, from the use of chains for lifting the bars. The second is prolonged exposure to sunlight, as ultraviolet light tends to make the coating become brittle. Bars are tied together to form reinforcement cages using plastic coated wire, again to avoid damage. Any areas of damage observed prior to casting are touched up using the epoxy repair compound.

There is a possibility that the coating may be damaged locally during the casting process. Care should be taken to avoid contact between vibration pokers and the reinforcement; to minimise the problem, some equipment manufacturers now offer soft rubber covering for the vibration equipment.

1.5.4   Design implications

The only aspect of design that will be significantly changed by the introduction of fusion-bonded epoxy-coated reinforcement is the bond. This will affect the anchorage, or development, length and the length of laps or splices. The American building code ACI 318 [12] requires the length to be increased by 50% when the cover is less than three bar diameters and the clear spacing between bars is less than six bar diameters. In all other conditions, only a 20% increase is required. Although BS8007 [28] suggests that it may be used to improve durability, British codes do not give any guidance on design with fusion-bonded epoxy-coated reinforcement.

It should be possible to reduce the thickness of the cover when using epoxy-coated reinforcement. However, there may be limited scope for doing this as the minimum cover is also dictated by the considerations of bond and by the size of the aggregate being used. Until guidance is available in codes and standards, it is probably wisest to leave the cover for durability requirements unaltered and consider the coating as an additional safeguard. As indicated earlier, the Japan Society of Civil Engineers Specification [16] would appear to be the only document at present that specifically allows reduced covers.

1.5.5 Applications

The most significant use of fusion-bonded epoxy-coated reinforcement to date has been in bridges. In the United Kingdom, the first major use in a bridge was in the parapets of a 1.5-km long viaduct in Cardiff. It has recently been used for two railway bridges, in Worcester and Reading, for all areas of concrete likely to be contaminated by salt spray from the road below. In addition, it has been used for marine structures and for coastal protection. A recent example was the sea wall at Hunstanton (see Figure 1.4), where it was included in the precast concrete units.

Figure 1.4 Sea wall at Hunstanton, incorporating epoxy-coated reinforcement.

In the United States and Canada, fusion-bonded epoxy-coated re-inforcement has been widely used in bridges, including major structures such as the Florida Sunshine Bridge and the Hood Canal floating bridge. It is estimated that more than 330 000 tonnes of coated reinforcement is used per annum, supplied by about 30 manufacturers.

Fusion-bonded epoxy-coated reinforcement has established a significant market in the Middle East. Here the hot, humid climate, often combined with a high level of chlorides in the local aggregates has led to very poor durability of reinforced concrete structures in the past. Coated reinforce-ment is thus an obvious method for improving durability. A number of plants are in operation, or are planned in the area.

1.5.6   Costs

The application of fusion-bonded epoxy-coating roughly doubles the price of the reinforcing steel as delivered to site. However, the true cost should be considered in terms of that of the total project cost. For example, for the railway bridge at Reading mentioned above, the use of coated steel was reported to have added only 1.5% to the overall cost. This agrees with cost comparisons by Wills [29] who studied two typical bridges, each with a span of nearly 30 m. He concluded that the use of epoxy-coated reinforcement would add, on average, 5% to the cost of the super-structure or between 1 and 2% to the total cost of the bridge.

1.5.7   Limitations

It has been suggested that localised damage to the coating can lead to concentrated corrosion damage. However, it has been shown in tests that little corrosion takes place at the localised points and that red rust does not form [30].

A further limitation is that coated reinforcement cannot be welded without destroying the coating. Systems have been developed for coating complete fabricated cages; these are discussed in the following section.

Doubts have been expressed concerning the performance of epoxy coated reinforcing bars in fire. Lin et al. [31] carried out a standard fire test on a $4.3 \times 5.5\,m^2$ slab reinforced with epoxy coated bars. The results were compared with those from a similar slab with uncoated reinforce-ment. The cover to the reinforcement was 20 mm. After 4 h, there was no significant difference between the two slabs although the one with the coated bars showed a somewhat greater central deflection. As part of the study, Lin et al. [32] also considered the effect of fire on the bond capacity of epoxy-coated bars. They showed that greater slip occurred, due to the softening of the coating. The load capacity for the coated bars was found to be about 20% less than for uncoated bars, a similar difference to that at ambient conditions. In other words, the coating would have no effect on the safety of a structure in a fire.

1.5.8   Post fabrication coating

One of the criteria for the choice of a suitable material and coating thickness is its ability to remain bonded to the reinforcement when the bar is bent. A thicker, less flexible coating may be more suitable from the point of view of long-term durability. There may be situations in which a welded reinforcement cage is desirable, but obviously the process of welding will destroy the coating. Hence, processes have been developed for coating a complete fabricated cage. An example was the reinforcement cages for the precast tunnel linings for the Great Belt Tunnel in Denmark [33]. The coating plant was built as part of the precasting factory for the tunnel linings, with a total of 61 000 reinforcement cages being required in all.

Eight precast segments went together to form a complete tunnel lining ring which had an internal diameter of 7.7 m. Each fully welded reinforcement cage consisted of an outer and an inner skin, connected by mats of steel running through the thickness. The coating process was similar to that for straight bars, i.e. (1) grit-blasting to clean the surface; (2) heating; (3) coating by fluidised dipping; (4) curing and inspection.

One limitation with the process is that larger diameter bars tend to gather a thicker coating than smaller ones, because of their different rates of cooling. However, the authors considered that the resulting cage would be of higher quality than the equivalent one made from bent and tied precoated bars and there would be less risk of mechanical damage during handling and placing in the mould. It was estimated that the coating added between 5 and 10% to the cost of the finished cage.

1.5.9   Coated prestressing strand

Epoxy-coated prestressing strand was developed by the Florida Wire and Cable Company in the United States, particularly for use in precast pretensioned members. Initial trials showed that transfer lengths were about 15% higher with coated strand than with uncoated [34]. Later work by Cousins et al. [35] included the effect of a grit impregnated in the epoxy to increase the bond. Tests were carried out under both static and fatigue loading. The authors concluded that the transfer length of grit-impregnated epoxy-coated strand was actually less than that of uncoated strand. With time, the transfer length of coated strand increased slightly more than uncoated strand.

1.6   Galvanised reinforcement

1.6.1   Introduction

Galvanised reinforcement consists of standard black bar, hot dipped in molten zinc. This process forms a coating which is metallurgically bonded to the surface of the parent metal. The surface of the zinc reacts with calcium hydroxide in the concrete to form a passive layer, preventing corrosion. It has been used in bridges and similar high value structures for the last 20 years or so in the United States and parts of Europe.

1.6.2   Production and use

Standard reinforcing bars, either in straight lengths or in prefabricated cages, are dipped in molten zinc. Processing straight lengths is less expensive than processing cages and is easier to control, giving more uniform coatings. However, if the straight lengths are subsequently to be cut and bent, they will have exposed ends and there is a danger of the coating flaking off tight radius bends. In this case, dipping the complete cage may be a better solution. However, it is a more difficult process to control as uneven coatings can form and it is more expensive.

From the designer's viewpoint, no changes are required when using galvanised reinforcement. It is assumed that the zinc coating is sufficiently stiff such that the bond characteristics of the bar are unaltered and hence anchorage and lap lengths are unchanged. Galvanised reinforcement is included in ACI 318 [12] but no special provisions are made.

1.6.3   Durability

Work on corrosion-resisting reinforcing steels carried out by Treadaway et al. [5] included some specimens containing galvanised reinforcement. (A summary of the programme is given in Section 1.7.3.) They found that galvanised steel showed significantly improved durability, when compared with standard high yield reinforcement, in carbonated concrete or in the presence of low levels of chloride. For high levels of chloride (up to 5% by weight of cement), the onset of corrosion was delayed, but only marginally.

1.6.4   Applications

The longest experience of use is in Bermuda where galvanised reinforcement was first specified in the 1930s because of the level of chloride in the local coral aggregates and the fact that concrete was often made using

Figure 1.5 The GLC Building, London.

seawater. Standard black reinforcement has often corroded within 2 or 3 years while galvanised bar has performed significantly better.

The most impressive applications for galvanised reinforcement have been where architects have used it for thin structural elements or cladding panels. It was used in the Sydney Opera House in the 1960s, for the shell roof structures. In London, 1000 tonnes of galvanised reinforcement were used in the most exposed parts of the National Theatre and it was also used for the cladding panels of the GLC Building, shown in Figure 1.5.

A significant use of galvanised reinforcement has been in bridges. In France and the United States, a number of bridges have been in service for more than 15 years and are apparently performing satisfactorily.

1.7 Stainless steel reinforcement

1.7.1 Introduction

Stainless steel is the name given to a family of corrosion resistant steels containing a minimum of 12% chromium. On contact with air, the chromium forms a thin oxide layer on the surface of the steel. This is passive and resists corrosion. The addition of other elements such as nickel and molybdenum enhances the passivity and thus improves the corrosion resistance. As the oxide layer is formed by the elements in the stainless steel, rather than being an applied coating, it is in fact self-repairing. Thus, if damage does occur during handling and fixing, the passive oxide layer rapidly reforms and the corrosion resistance is not affected.

A range of stainless steels with different mechanical properties and degrees of corrosion resistance are available. The austenitic group of steels have superior corrosion resistance and are the most suitable for use in reinforced concrete.

1.7.2 Availability and use

In the United Kingdom, stainless steel reinforcing bars are covered by BS6744 [36]. Plain and ribbed bars are available in the same range of standard sizes as normal reinforcement, with similar characteristic strengths. The bond characteristics are also similar. Hence, no design or detailing changes are required when replacing normal carbon steel by stainless steel. However, physical contact between stainless steel reinforcement and other embedded metal, for example cast-in sockets or normal carbon steel reinforcing bars, should be prevented. This is to avoid the phenomenon of bimetallic corrosion in which the less noble metal (e.g. the carbon steel) will act as a sacrificial anode and will corrode more rapidly. Bimetallic corrosion would be of most significance in aggressive environments such as those where chlorides are present.

Stainless steel reinforcing bars have good mechanical properties at elevated temperatures. It is reported [37] that after exposure to heating to 600°C for 2 h, stainless steel bars showed a negligible reduction in tensile and yield stress while conventional reinforcement showed a significant drop. This suggests that concrete elements reinforced with stainless steel would behave better in a fire than would conventionally reinforced elements.

Austenitic stainless steels are less sensitive to low temperatures than conventional carbon reinforcement, with a ductile failure at temperatures as low as −196°C. They are therefore suitable for cryogenic applications.

1.7.3 Durability

Treadaway et al. [5] reported on a 10-year programme of work, carried out at the Building Research Establishment, which compared the behaviour of stainless steels and galvanised reinforcement with conventional high yield steel. Bars were cast into concrete specimens, using two different basic concrete mixes with five different amounts of chloride added and two different covers to the reinforcement. In all, 550 specimens were made and exposed in East London, mostly for the full 10 years, although some examinations were carried out during the period. At the end of the period, the bars were broken out of the concrete, cleaned then weighed to assess the loss due to corrosion. The authors concluded that

Figure 1.6 Stainless steel reinforcement being used in Sydney.

"all the austenitic stainless steels showed very high corrosion resistance in all the environments tested. No serious corrosion was encountered on any of the austenitic bars." Flint and Cox [38] studied the behaviour of austenitic stainless steel embedded in concrete when subjected to sea-water. Specimens were located near Portsmouth either totally submerged or else in the splash zone. After a total of 12 years, corrosion of the exposed stainless steel was localised, not extensive, and affected neither the strength nor the ductility of the specimens.

1.7.4 Applications

Stainless steel reinforcement has been used in a number of structures in aggressive environments. These include bridge decks in the United States, the waterside promenade in front of the Sydney Opera House, as shown in Figure 1.6, and precast facing units for a property on the coast at Scarborough. In addition to their non-corroding properties, austenitic stainless steels are virtually non-magnetic. Hence, they are used near sensitive electronic equipment where a neutral magnetic environment is required.

1.7.5 Costs

Stainless steel reinforcement is significantly more expensive than standard bar, possibly seven or eight times as much for the material as delivered to site. Of course, fixing costs will reduce the differential somewhat. A more realistic comparison is to consider the increase in the total cost of the structure. Comparisons by Wills [29] suggest that the use of stainless steel in areas subjected to salt or salt spray would add about 40% to the cost of the superstructure but only about 10% to the cost of the complete bridge.

1.7.6   Stainless steel coated bar

Because of the high cost of solid stainless steel reinforcing bars when compared to normal bars, attempts have been made to develop a bar with just a stainless surface. One such was 'Staiclad' in which a stainless steel coating was wrapped round the bar. The product proved to be difficult and costly to produce and is no longer on the market.

1.8   Non-ferrous reinforcement

1.8.1   Artificial fibres

There are currently a number of manufacturers who are developing non-ferrous reinforcement as an alternative to the conventional steel in traditional structures. Currently a range of man-made fibres are used, the most common being glass, carbon and aramid. The materials are being developed rapidly at present, but typical stress–strain diagrams are shown in Figure 1.7. It may be seen that aramid, glass and carbon fibres all have ultimate stresses well in excess of that of reinforcing steel. They have elastic moduli that range from about 35% of that of steel to perhaps 50% stiffer. One significant difference between these high grade fibres and steel is that the stress–strain curve does not show any sign of plasticity at high stresses. This could be a limiting factor in some design situations.

The fibres are used either in the form of ropes or combined with suitable resins to form rods. Because of their relatively low elastic modulus, they have generally been used for prestressing. Examples include a number of bridges in Germany using glass-fibre tendons, precast units in the Netherlands and unbonded structures in the United Kingdom. These are discussed in Chapters 6, 8 and 5, respectively. However, some products are being used as unstressed reinforcement. They consist of continuous glass or carbon fibres set in a resin to form reinforcement rods or grids. To date they have been used in a variety of concrete structures, generally in highly corrosive situations or ones in which stray electrical currents are undesirable. Examples include tunnel linings in Japan and beneath sensitive hospital equipment in the United States. Further details are given in Chapters 2 and 3. An interesting development has been the use of fibre composites for the repair and strengthening of structures, the

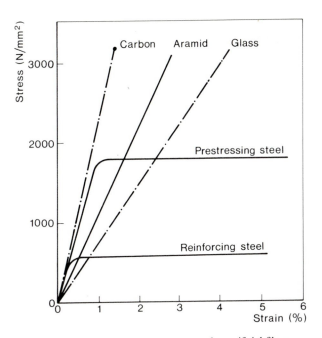

Figure 1.7 Typical stress–strain curves for artificial fibres.

material being bonded directly to the surface of the concrete. Chapter 7 describes applications using carbon fibres.

Properties of the various products currently available are compared in the Appendix.

1.8.2 Manufacturing process for fibre composites

Currently the most suitable manufacturing process for resin-fibre composites would appear to be pultrusion. This is a simple process that may be used to form a wide range of structural shapes, such as rods, flat strips, angles, etc., using different resins and fibres. The fibres may be all of one type or else a combination of materials.

For the pultrusion process, fibres are supplied either in the form of continuous rovings or mats, or as a combination of the two. They are drawn off in a carefully controlled pattern and then through a resin bath, which impregnates the fibre bundle. The excess resin is wiped off and the fibres are pulled through a die, which consolidates the resin-fibre combination and forms the desired shape. The die is heated, setting and curing the resin. The completed composite section is drawn off by means of suitable reciprocating clamps or a continuous tension device. Generally the sections are cut into suitable lengths for handling though flexible sections can be coiled.

The process enables a high percentage of fibre to be incorporated in the cross-section, typically about 70%. One limitation, as far as the reinforcement of concrete is concerned, is that pultruded sections have a smooth surface, which would not give sufficient bond. A secondary process is required to change the surface profile.

There are other systems available in which a continuous band of resin and fibres is laid down on a suitable base mould. This allows the production of two-dimensional grids which can be used for reinforcement; here the strips at right angles give sufficient bond and the smoothness of the resin surface is no longer a problem.

1.8.3   Polymer grids

While most developments of non-metallic materials have been with fibres, polymer grids are also being used. Details of the manufacturing process are given in Chapter 4. Briefly, sheets of polymer are punched with a regular array of holes. The sheet is then stretched in two directions under controlled temperature and strain conditions. During this process, the long chain molecules, which are randomly orientated within the original sheet, are aligned along the direction of stretch. This results in an increase in the tensile strength to produce biaxial grids which are primarily used in reinforced soil applications. The grids are inert and have no solvents at ambient temperature. They are therefore suitable for use as a reinforcement in aggressive environments. Some applications are discussed in Chapter 4.

1.9   Design with non-ferrous reinforcement

1.9.1   Introduction

In the United Kingdom, the Department of Transport has commissioned an outline design for a 20-m highway bridge with non-ferrous reinforcement and prestressing throughout. Before such materials can become more generally accepted, it will be necessary to develop suitable design approaches. Current Codes of Practice for reinforced and prestressed concrete are based on practical testing and experience of conventional steel. If the steel is replaced by non-ferrous material, the present design approaches may need to be modified to take account of the different material properties. Subsequent chapters consider specific aspects of design applicable to the particular material being discussed. The aim of this section is to review the design aspects in which there are likely to be significant differences and to identify areas in which research is required before alternative reinforcement can be used with full confidence.

1.9.2 Design criteria

Design is carried out at the ultimate limit state, to ensure that the structure has adequate strength, and at the serviceability limit state to ensure that its behaviour is satisfactory during the intended design life. The significant design aspects are as follows:

1. Ultimate: bending; shear; torsion; fatigue.
2. Serviceability: deflections; crack widths; cover to reinforcement; fire.

Each is considered in turn. In addition, there are aspects of detailing that could influence the behaviour at both ultimate and service loads.

1.9.3 Bending

Failure of a reinforced or prestressed concrete member in bending is due to yield/rupture of the reinforcement or crushing of the concrete. Knowing the stress–strain behaviour of the alternative material, the bending response could be easily determined from first principles. However, there will be two significant differences that must be allowed for. The first is that the alternative materials have a straight elastic stress–strain response throughout and show no significant yield before failure. This might necessitate higher partial safety factors. The bridge code, BS5400 [11], requires a 15% greater bending moment capacity for sections that fail in a brittle manner. The second difference is that, under sustained load, the alternative materials will creep to failure (known as stress rupture). However, for a given design life, the reinforcement stress to cause failure can be determined and this value used as the maximum permitted stress under service load conditions. The significance of this limitation could be determined readily by parametric studies.

1.9.4 Shear and torsion

The shear failure of beams, or the punching of slabs, occurs with little or no warning. Current design methods are empirical. For reinforced concrete, these are based on both the strength of the concrete and the amount of tensile reinforcement present at the section under consideration. Extensive testing would be required to determine similar empirical design approaches for alternative reinforcement.

For prestressed concrete, BS8110 and BS5400 [8, 11] identify two types of failure. The first, shear of sections uncracked in bending is based on the concept that the principal tensile stress should not exceed the tensile strength of the concrete. Calculations are carried out at the centroid and depend on the stress due to prestress. This latter would not be influenced by changing from steel to an alternative material. The second type of shear is an empirical expression based on extensive testing [39]. Again further testing would be required to develop comprehensive design rules for alternative materials.

Once the shear capacity of the concrete section is exceeded, shear reinforcement must be provided. In both BS8110 and BS5400, in line with most other structural codes, the total shear capacity is taken as the capacity of the concrete plus that of the links. This is based on the assumption of a controlled shear crack across which a significant proportion of the uncracked shear capacity can still be transferred. It is unlikely that this will be true with alternative reinforcement as the lower elasticity will lead to much wider cracks. It may be necessary to revert to earlier design approaches in which, once the capacity of the concrete had been exceeded, all the shear had to be carried on shear reinforcement.

The maximum shear that can be carried by a section is dictated by the crushing of the compressive strut forming part of the truss with the links. Both BS5400 and BS8110 use an empirical approach, which is a safe lower bound to test evidence. It is likely that the change to non-ferrous reinforcement will have little significant difference on the value but some testing would be required to confirm this view.

For most structures, torsion is not a significant part of the design process. Where torsional design is required, the current approach is to provide torsional and additional longitudinal reinforcement, based on the section geometry and the characteristic strength of the reinforcement. Thus, it should be possible to substitute alternative reinforcement directly, but it might be necessary to carry out limited tests to confirm this.

1.9.5 Fatigue

Fatigue will only be of significance in reinforced or prestressed concrete in slender members subjected to high dynamic loading. Failure will generally be in the reinforcement. Thus, a knowledge of the fatigue behaviour of the alternative reinforcement will indicate the fatigue life of the structure.

1.9.6 Deflections and cracking

For an uncracked section, the deflection will be a function of the gross concrete cross-section only and hence not influenced by the type of reinforcement. For the fully cracked situation, the deflection will depend on the strain in the reinforcement, which can be determined if the stress–

strain response is known. For a 'real' situation, i.e. between uncracked and fully cracked, the actual deflection depends on the distribution of cracking which will be influenced by the bond between the reinforcement and the concrete. It is likely that deflection, rather than strength, would be the governing criterion for many structures reinforced with non-ferrous material.

Crack widths have traditionally been limited because it was thought that they influenced the corrosion of the reinforcement. It is now generally accepted that crack widths have little or no significant influence on the long-term durability of reinforced concrete structures. Hence, the major reason for controlling crack widths would appear to be from the point of view of aesthetics. For many structures, viewed from some distance away, realistic limiting widths with alternative reinforcement could be greatly in excess of those currently specified. However, assuming that some crack width calculation would still be required, experimental work would be required to check the validity of the current design methods or to propose others.

1.9.7 Cover to reinforcement

The cover concrete is primarily to provide protection to the reinforcement, either from the point of view of durability or from considerations of fire. In addition, a minimum cover, depending on the bar size and the maximum size of aggregate, is required to ensure adequate bond between the bar and the concrete. This latter will still be required for the alternative reinforcements but additional cover for durability will no longer be necessary.

Limited test data available suggest that, with the covers currently specified for fire resistance, some alternative materials will behave satisfactorily. However, more information is required on the behaviour of both the fibres themselves and the fibre/resin composites at elevated temperatures.

1.9.8 Detailing

There are several areas of detailing that require consideration including end anchorage, laps, bearing inside bends, etc., but none is likely to have a significant influence on design. However, because most alternative materials are currently produced in straight lengths which cannot be bent, hooks and bends can only be provided as specials. This would have an effect on detailing practice. The alternative would be to develop standard two- or three-dimensional grids, which would provide both tensile and shear reinforcement.

At present, the requirements for minimum amounts of reinforcement, which are provided to avoid large cracks and to ensure a ductile behaviour, are largely empirical. Hence, they need careful study.

1.9.9   Priorities for research

In the light of the above, it may be seen that significant research and development work is required before non-ferrous reinforcement can be fully incorporated into codes and standards. The priorities are as follows:

1. High priority: shear of beams; punching of slabs; fire resistance.
2. Medium priority: crack distribution; deflection.
3. Low priority: torsion; detailing.

The following chapters show how many of the design aspects have been addressed in particular applications, using particular products.

1.10   Summary and conclusions

This chapter has dealt with the mechanism of corrosion and the approaches currently adopted by codes and standards, to ensure durability. These rely on an adequate thickness of good quality concrete, often combined with some limitation of the surface crack width. However, this approach has often proved inadequate, particularly in highly aggressive environments. The designer has at his disposal a range of special products that enhance the behaviour of the concrete or protect the steel from corrosion. Normal carbon steel may be replaced by stainless steel. All of these may be described as 20th century solutions. Now reinforced and prestressed concrete structures are set to move into the 21st century with a range of new non-ferrous materials to replace the steel.

References

1. Woodward, R.J. and Williams, F.W., Collapse of Ynys-y-Gwas bridge, West Glamorgan. *Proc. Inst. Civil Eng.* **84** (1988) 635–669.
2. Brown, J.H., The performance of concrete in practice: a field study of highway bridges, Transport and Road Research Laboratory, Contractor Report 43, 1987.
3. Leeming, M.B., Concrete in the oceans: coordinating report on the whole programme, Offshore Technology Report OTH 87,248, HMSO, 1989.
4. Read, J.A., FBECR, The need for correct specification and quality control. *Concrete* **23**(8) (1989).
5. Treadaway, K.W.J., Cox, R.N. and Brown, B.L., Durability of corrosion resisting steels in concrete. *Proc. Inst. Civil Eng.* **86** (1989) 305–331.
6. Beeby, A.W., Cracking and corrosion, concrete in the oceans, Technical Report No. 1, CIRIA/UEG, Cement and Concrete Association, Department of Energy, 1978, p. 77.
7. BS5328, Concrete, British Standards Institution, London, 1990.

8. BS8110, Structural use of concrete, British Standards Institution, London, 1985.
9. Rowe, R.E. *et al.*, *Handbook to British Standard BS8110:1985. Structural Use of Concrete*, Palladian Publications, London, 1987.
10. BS6349, Maritime structures, British Standards Institution, London, 1984.
11. BS5400, Steel, concrete and composite bridges, Part 4, Code of practice for design of concrete bridges, British Standards Institution, London, 1990.
12. ACI 318-89, Building code requirements for reinforced concrete, American Concrete Institute, Detroit, 1989.
13. ACI 357.IR-91, State-of-the-art report on offshore concrete structures for the Arctic, American Concrete Institute, Detroit, 1991.
14. ENV 1992-1-1: Part 1, Design of concrete structures, Part 1, General rules and rules for buildings. Final draft, European Committee for Standardization, Brussels, 1991.
15. ENV206, Concrete—Performance, production, placing and compliance criteria, European Committee for Standardization, Brussels, 1990.
16. Japan Society of Civil Engineers, Standard Specification for design and construction of concrete structures, Part 1, Design, Tokyo, English translation 1986.
17. ACI Committee 212, Admixtures for concrete and guide for use of admixtures in concrete. *Concrete Int.* **May** (1981) 24–25.
18. Elkem Materials, Elkem microsilica for concrete impermeability, Data sheet 4.1.301E, 1991.
19. Berke, N.S. *et al.*, Protection against chloride-induced corrosion. *Concrete Int.* **December** (1988) 45–55.
20. Alonso, C. and Andrade, C., Effect of nitrate as a corrosion inhibitor in contaminated and chloride-free carbonated mortars. *Am. Concrete Inst. Mater. J.* **March–April** (1990) 130–137.
21. Perenchio, W.F., Durability of concrete treated with silanes. *Concrete Int.* **November** (1988) 32–40.
22. Alexander, D., Silane monopoly under threat. *New Civil Eng.* **23 May** (1991) 5.
23. Weyers, R.E. and Cady, P.D., Cathodic protection of concrete bridge decks, *J. Am. Concrete Inst.* **November–December** (1984) 618–622.
24. Robbins, J., DTp combats corrosion with cathodic spray. *New Civil Eng.* **15 August** (1991) 8.
25. Clifton, J.R., Beeghly, H.F. and Mathey, R.G., Final report, Non metallic coatings for concrete reinforcing bars, Federal Highways Administration, Report RD-18, 1974.
26. BS7295, Fusion bonded epoxy coated carbon steel bars for the reinforcement of concrete, British Standards Institution, London, 1990.
27. ASTM A775, Standard specification for epoxy coated reinforcing bars, American Society for Testing and Materials, Philadelphia.
28. BS8007, Design of concrete structures for retaining aqueous liquids, British Standards Institution, London, 1987.
29. Wills, J., Epoxy coated reinforcement in bridge decks, Supplementary Report 667, Transport and Road Research Laboratory.
30. British Board of Agreement, Ebar, fusion bonded epoxy coated reinforcing bars, Certificate No. 90/2425, 1990.
31. Lin, T.D. *et al.*, Fire test of concrete slab reinforced with epoxy-coated bars. *Am. Concrete Inst. Struct. J.* **March–April** (1989) 156–162.
32. Lin, T.D. *et al.*, Pull-out tests of epoxy-coated bars at high temperatures. *Am. Concrete Inst. Mater. J.* **November–December** (1988) 544–550.
33. Ecob, C.R. *et al.*, Epoxy coated reinforcement cages in precast concrete segmented linings—durability, *SCI 3rd Int. Conf. on Corrosion of Reinforcement in Concrete Construction*, Warwickshire (1990).
34. Dorston, V. *et al.*, Epoxy coated seven-wire strand for prestressed concrete. *J. Prestressed Concrete Inst.* **29**(4) (1984) 120–129.
35. Cousins, E.T. *et al.*, Transfer length of epoxy-coated prestressing strand. *Am. Concrete Inst. Mater. J.* **May–June** (1990) 193–203.
36. BS6744, Austenitic stainless steel bars for the reinforcement of concrete, British Standards Institution, London, 1986.

37. Stainless steel reinforcing bar, Technical Leaflet, George Clark (Sheffield) Ltd., Sheffield.
38. Flint, G.N. and Cox, R.N., The resistance of stainless steel partly embedded in concrete to corrosion by sea water. *Mag. Concrete Res.* **40**(142) (1988) 13–27.
39. Reynolds, G.C. *et al.*, Shear provisions for prestressed concrete in the Unified Code, CP110:1972, Report 42.500, Cement and Concrete Association, London, 1974.

Appendix

This appendix compares some of the properties of the alternative material products discussed in the book. Also included are the properties of standard reinforcing and prestressing steels by way of comparison, as well as those of some alternative materials not included in the text. The ultimate stresses and the moduli of elasticity are effective values, based on the full cross-section for resin-fibre composites.

Table A.1 Comparison of materials in terms of ultimate stress

Material	Ultimate stress (N/mm²)
Parafil F and G	1930
Tokyo Rope (carbon fibre strand)[a]	1770
Prestressing steel	1600–1860
Arapree (circular cross-section)	1700
Polystal	1670
JITEK (glass or carbon fibre composite)[a]	1000–1600
Arapree (rectangular cross-section)	990–1190
Prestressing bar	1000
Kodiak	760–1030
Nefmac	625–700
High yield reinforcing steel	460
Tensar	180–410

[a] Not described in text.

Table A.2 Comparison of materials in terms of modulus of elasticity

Material	Modulus of elasticity (kN/mm²)
Steel reinforcement, prestressing bar and wire	200
Steel prestressing strand	175–195
Tokyo Rope (carbon fibre strand)[a]	137
Parafil G	127
Parafil F	80
Arapree (circular cross-section)	70
Polystal	51
Arapree (rectangular cross-section)	50
Kodiak	45–50
JITEK (glass)[a]	35–50
Nefmac	35–45
Tensar	4–7

[a] Not described in text.

Table A.3 Comparison of residual strengths at elevated temperatures

Materials	Residual strength (%)			
	150°C	200°C	300°C	400°C
Prestressing strand	100	90	70	50
Reinforcement	100	100	100	80
Arapree[a]	100	95	85	55
Polystal	95	90	80	55
FRP rebar	90			

[a] Heated for 0.5 h.

2 Glass-fiber reinforcing bars

M.R. EHSANI

2.1 Introduction

Corrosion of steel reinforcement is a major concern in concrete construction located in aggressive environments such as coastal and marine structures, chemical plants, water and wastewater treatment facilities and bridges, especially when de-icing chemicals are used. Rust from corroded rebars occupies a larger volume than the iron from which it is formed. This results in large internal pressures which lead to cracking and spalling of concrete and ultimately the failure of the structure. Several methods for controlling the corrosion process have been developed which include increasing the concrete cover, improving the permeability of concrete by the use of additives and admixtures, cathodic protection and epoxy-coating of the reinforcing bars. The last has been used extensively in North America since it was first introduced in a Pennsylvania bridge in 1973 [1]. However, recent reports on corrosion of epoxy-coated reinforcing bars in bridges in Florida Keys have lowered the expectations of the engineering community on the long-term performance of such coatings [2]. In fact, these findings have resulted in an undertaking by the US Federal Highway Administration to concentrate on the development of new materials or coatings for reinforcing bars. A completely different approach would be to use materials that are highly corrosion resistant, such as reinforcing bars constructed of composite materials.

Since the 1930s, glass has been considered as a potential substitute for steel in reinforcing or prestressing concrete structures. However, the initial studies noted problems with anchorage and surface protection of bars, bonding of glass fibers to concrete, etc. [3–6]. There was a period of some 20 years where research in this field remained dormant for the most part. Due to the increased use of composite materials in other applications such as aerospace, chemical, shipbuilding, etc. since the 1970s, great advances have been made in this field. Therefore, many of the initial shortcomings have been overcome, resulting in renewed interest in the use of these materials in civil engineering.

While in some fields such as the aerospace industry, the use of advanced composite materials has mainly focused on graphite, aramid, or other costly high performance fibers, the most suitable fiber for civil engineering applications is glass. As discussed in this chapter, the physical and mechanical properties of glass, along with its low cost make it an ideal material for civil engineering construction where usually large volumes of materials are needed. Consequently, the focus of this chapter is on glass-fiber reinforced plastic (GFRP) rebars.

2.2 Manufacturing, shapes and sizes

2.2.1 Introduction

Fiber composites, in general, are constructed of strands of one or more types of reinforcing fibers which are bonded together through a resin or matrix system. Depending on the application, the role of the matrix can be either as a binder which contains the major structural elements (i.e. fibers) and transfers the load between them, or as the primary load carrying phase which is reinforced with fibers. The former approach is more common since the matrices used in most composites are relatively soft thermosetting plastics such as epoxies, polyesters, or phenolics.

Until recently, the use of composite materials was limited to fields where, due to the small volumes of material needed, the unit cost was not a significant factor in the selection process. Examples of such applications are the aerospace and defence industries where intricate high performance components have been constructed at high costs. For civil engineering applications, the use of such materials could not be justified. Fortunately, with recent developments in fibers, resins and manufacturing processes, the cost of fiber reinforced plastics (FRP) has been reduced to a point where they can compete with conventional materials such as steel.

There are several techniques for manufacturing of FRP composites, including hand lay-up, compression molding, pultrusion, injection molding, etc. [7]. Regardless of the manufacturing technique, there are three stages that are common to construction of FRP composites: (1) the primary processing of the constituent materials, i.e. resin and fibers; (2) the intermediate processing of these materials into required forms; and (3) the shaping and curing of the combined constituents into the final geometry.

2.2.2 Pultrusion process

Due to its fast speed of operation, good quality control and relatively low equipment cost, pultrusion is the method commonly utilized for manufacturing FRP composites used in the construction industry. Pultrusion is particularly suitable for manufacturing composite structures with continuous, constant cross-sectional profiles, e.g. reinforcing bars. In recent years, there has been a proliferation of companies worldwide which produce various pultruded structural shapes such as I-beams, angles, channels, plates, tubes, etc.

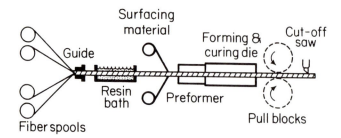

Figure 2.1 Pultrusion process.

The pultrusion process used for manufacturing FRP of composite rebars is similar to that shown in Figure 2.1. The process starts with several spools of fibers in the form of strands. Although glass is the fiber most commonly used, it is possible to combine different fibers at this stage to obtain the desired physical characteristics for the finished product. In some cases, the strands are first passed over a series of heaters to remove any moisture condensation from the fibers. The fibers are then pulled through a series of guides where they are formed into the desired shape. Next, the material is passed through a resin bath for impregnation. The resin is mixed with the necessary accelerators, catalysts, filler materials and other required additives.

During the last stage, the resin impregnated fibers are passed through the final heated dies where the excess resin is squeezed out of the rebar. If the resin system is a thermoset (e.g. epoxies and polyesters), the heated dies initiate the curing process. The dies also insure uniformity of finished dimensions and maintain the desired volume of resin in the rebar. As the completed rebar is pulled out of the production line, it can be cut with a saw to the desired length. Some manufacturers have proprietary techniques where, before the rebar is cured, a set of strands is helically wrapped around the rebar [8]. The deformations thus formed on the surface of the rebar improve its bond strength to concrete.

2.2.3 Available sizes

GFRP rebars are manufactured in a wide range of sizes. Unlike steel, there are no standards for the size of GFRP rebar. However, most manufacturers produce these bars in the same diameters as steel rebars. In the United States, for example, these bars are available in diameters ranging from 2/8 to 9/8 in (6–29 mm) [8]. This obviously simplifies the task of the designer since he can obtain the cross-sectional properties from existing tables for steel rebars.

Due to this lack of standards, the quality as well as the physical dimensions of GFRP rebars may vary widely from one manufacturer to another. This non-uniformity may appear in several ways. It is possible, for example, to find slight variations in diameters of the rebars of a given size manufactured in the same plant. Some rebars, especially the larger sizes, may have cross-sections that are slightly oval shaped rather than a true circle. When strands are wrapped in a helical pattern around the bar, the pitch distance may include inconsistencies. While it is true that the strength of GFRP rebars is primarily a function of the volume of fibers used in the longitudinal direction, the effect of some of these geometric variations on certain characteristics such as bond strength may be significant. There is no doubt that as the use of these materials increases, standards will be developed to ensure uniformity of production based on minimum performance requirements.

2.2.4 Available shapes

The smaller diameter bars can be force bowed much like a fishing rod and tied in place to maintain the curvature. Sharper bends, similar to the standard 90° or 180° hooks cannot be performed in the field. Attempts to bend a straight bar or straighten a bent rebar will result in very large stresses in the outermost fibers at those locations, leading to fracture of the bar. It is noted, however, that FRP rebars can be manufactured in any shape in the plant while the resin has not cured. Shapes such as closed rectangular stirrups or ties, spiral reinforcement for columns and 90° and 180° hooks are commonly manufactured by FRP rebar producers. Some of these are shown in Figure 2.2. However, due to the required labor, there is a cost premium associated with the manufacturing of these shapes.

The inability to bend the bars in the field may call for some advanced planning to order the required rebars before construction begins. In some cases, there are other ways to overcome this shortcoming. For example, for construction of a retaining wall with varying height, 90° hooks with identical dimensions can be lap spliced in the field with straight bars which are cut to various lengths to obtain the desired dimensions for reinforcement.

The use of thermoplastic resins in the construction of FRP rebars is also under investigation. Thermoplastic rebars, which are more expensive than thermosets, can overcome some of the shortcomings of thermosets, such as welding and reshaping in the field [9].

GFRP rebars can be cut easily in the field with portable saws using masonry blades or with hacksaws. Care must be taken to seal these locations in the field to prevent the penetration of moisture which could damage the rebars. Most manufacturers provide kits containing resins for sealing the cut ends.

Figure 2.2 Sample of reinforcing shapes constructed with GFRP rebars.

2.3 Mechanical properties

Behavior of GFRP rebars can be divided into two categories. The first category includes the mechanical properties of the rebar itself. These are very similar to the characteristics of any GFRP material and are covered in this chapter. The second group is the structural behavior of concrete members reinforced with GFRP rebars. These include such aspects as flexural, shear, axial, and bond strength, serviceability, etc., and are covered in the next chapter.

In discussions related to the properties of FRP rebars, the following points must be kept in mind. First, FRP rebars are strongly anisotropic, with the longitudinal axis being the strong axis. Second, unlike steel, the mechanical properties of FRPs vary significantly from one product to another. Factors such as volume and type of fiber and resin, fiber orientation, quality control during the manufacturing, etc., play a major role in the characteristics of the product. Furthermore, the mechanical properties of GFRPs, like all composites, are affected by such factors as the loading history and duration, temperature and moisture.

While for traditional construction materials, such as steel and concrete, standard tests have been established to determine their properties, the same cannot be said for composite materials. This is particularly true for civil engineering applications, where the use of composite materials is in its stage of infancy. It is therefore required that the exact loading conditions be determined in advance and the material characteristics corresponding to those conditions be obtained for design purposes.

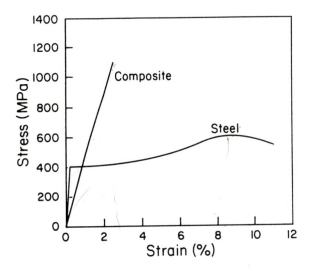

Figure 2.3 Stress versus strain relationship for steel and GFRP rebars.

2.3.1 Tensile strength

GFRP rebars reach their ultimate tensile strength without exhibiting any yielding of the material. A comparison of the properties of GFRP rebars and steel rebars is shown in Figure 2.3 and Table 2.1. The mechanical properties of GFRP rebars reported here are measured in the longitudinal (i.e. strong) direction. It is noted that although the tensile strength of GFRP rebars is slightly higher than that of steel, it is significantly higher than the yield strength of steel rebars.

Unlike steel, the tensile strength of GFRP rebars is a function of the bar diameter. Due to shear lag, the fibers located near the center of the bar cross-section are not subjected to as much stress as those fibers which are near the outer surface of the bar [10]. This phenomenon results in reduced strength and efficiency in larger diameter rebars. Table 2.2 lists the tensile strength of some of the different diameter rebars manufactured in the United States [8].

2.3.2 Tensile modulus of elasticity

As noted in Table 2.1, the modulus of elasticity of GFRP rebars is approximately 25% of that of steel. This results in a smaller transformed moment of inertia for flexural members reinforced with GFRP rebars. Consequently, unless the rebars are prestressed or post-tensioned, deflection may control the design of flexural members.

Table 2.1 Comparison of properties of steel and GFRP rebars

	Steel rebar	GFRP rebar
Tensile strength (MPa)	500–700	550–1500
Yield strength (MPa)	280–420	
Tensile modulus of elasticity (GPa)	200	41–55[a]
Compressive strength (MPa)	500–700	320–470
Compressive modulus of elasticity (GPa)	200	34–47
Coefficient of thermal expansion ($\times 10^{-6}$/°C)	11.7	10.0
Specific gravity	7.9	1.5–2.0

[a] For samples having tensile strength ranging from 550 to 900 MPa.

Table 2.2 Sample properties for GFRP rebars

Bar no.	Diameter (mm)	Cross-sectional area (mm^2)	Tensile strength (MPa)
3	9	71	900
4	13	127	740
5	16	198	655
6	19	285	620
7	22	388	585
8	25	507	550

2.3.3 Compressive strength

GFRP rebars are weaker in compression than in tension. However, the compressive strength of GFRPs is not a primary concern for most applications. The compressive strength also depends on whether the rebar is smooth or ribbed. Studies conducted by Wu report a compressive strength ranging from 320 to 470 MPa for rebars having a tensile strength in the range of 550–900 MPa [11]. Higher compressive strengths are expected for bars with higher tensile strength.

2.3.4 Compressive modulus of elasticity

Unlike the tensile stiffness, the compressive stiffness of GFRP rebars varies with rebar size, type, quality control in manufacturing and the length to diameter ratio of the specimens. The compressive stiffness for GFRP rebars is smaller than the tensile modulus of elasticity. Based on tests of samples containing 55–60% volume fraction of continuous E-glass fibers in a matrix of vinylester or isophthalic resin, a modulus of 36–47 GPa has been reported [11]. Another manufacturer reports the compressive modulus at 34.5 GPa which is approximately 77% of the tensile modulus for the product [12].

2.3.5 Shear strength

The shear strength of composites is, in general, very low. GFRP rebars, for example, can be cut very easily in the direction perpendicular to the longitudinal axis with ordinary saws. This shortcoming can be overcome in most cases by orienting the rebars such that they will resist the applied loads through axial tension.

2.3.6 Factors affecting mechanical properties

The mechanical properties of composites are dependent upon many factors such as load duration and history, temperature and moisture. These factors are interdependent and consequently it is not possible to determine the effect of each one in isolation while the others are held constant.

2.3.6.1 Creep. Fibers such as graphite and glass have excellent resistance to creep, while the same is not true for most resins. Therefore, the orientation and volume of fibers have a significant influence on the creep performance of composites. One study reports that for a high quality GFRP rebar, the additional strains caused by creep were estimated to be only 3% of the initial elastic strains.

Under adverse loading and environmental conditions, composites subjected to the action of a constant load may suddenly fail after a time, referred to as the endurance time. This phenomenon, known as creep rupture, exists for all structural materials including steel. However, for prestressing strands for example, this is not of concern; steel can endure typical tensile loads, which are about 75% of the ultimate strength, indefinitely without any loss of strength or fracture. As the ratio of the sustained tensile force to the short-term strength of the FRP increases, the endurance time decreases. Creep tests were conducted in Germany on GFRP composites with various cross-sections at 20°C in air. These studies indicate that stress rupture diminishes if the sustained loads are limited to 60% of the short-term strength of the sample [13].

The above limit on stresses is of little concern for most reinforced concrete structures where the sustained stresses on the reinforcement are significantly below 60%. However, it does require special attention in applications of GFRP as prestressing tendons. It must be noted that other factors such as moisture also impair creep performance and may result in shorter endurance time.

2.3.6.2 Fatigue. Composites reinforced with long fibers exhibit good fatigue resistance. Most of the research in this regard has been on high performance fibers, such as graphite, which are subjected to large cycles of loading in aerospace applications. In tests where the loading was repeated for 10 million cycles, it was concluded that graphite-epoxy composites have better fatigue strength than steel, while the fatigue strength of glass composites is lower than steel [14]. In another investigation, GFRP rods constructed for prestressing applications were subjected to repeated cyclic loading with a maximum load of 500 MPa and a load range of 350 MPa. The rods could stand more than 4 million cycles of loading before the failure initiated at their anchorage zone [15].

2.3.6.3 Moisture. Excessive absorption of water in composites could result in significant loss of strength and stiffness. Water absorption results in changes in the properties of the resin and could cause swelling and warping in composites. It is therefore imperative that mechanical properties of the composites be determined for the same environmental conditions where the material is to be utilized. There are, however, resins which are moisture-resistant and may be used when the structure is expected to be wet at all times. In cold regions, the effect of freeze–thaw cycles of the moisture must also be considered.

2.3.6.4 Fire. Many composites have good to excellent properties at elevated temperatures. Most composites do not start to burn easily. The effect of the high temperature is more severe on the resin matrix than the fiber. Resins contain large amounts of carbon, nitrogen and hydrogen which are flammable and research is continuing on the development of more fire-resistant resins [14]. Tests conducted in Germany have shown that when composite rods of E-glass have been stressed to 50% of their tensile strength, after 0.5 h of exposure to a temperature of 300°C, the bars can maintain 85% of their room-temperature strength [15]. While this performance is better than that of prestressing steel, the decrease in strength increases at higher temperatures and approaches that of steel. At 500°C, for example, GFRP rebars maintain 50% of their ambient temperature strength.

The problem of fire for concrete members reinforced with GFRP rebars is different from that of composite materials subjected to direct fire. In this case, the concrete will serve as a barrier to protect the GFRP rebars from direct contact with flames. However, as the temperature in the interior of the member increases, the mechanical properties of the GFRP rebars may change significantly. It is therefore recommended that the user obtain information on the performance of the particular GFRP rebar and resin system at elevated temperatures when potential for fire exists.

2.3.6.5 Ultraviolet rays. Composites can be damaged by ultraviolet rays present in sunlight. These rays cause chemical reactions in the polymeric matrices which can lead to a degradation of their properties. Although the problem can be solved with the introduction of appropriate additives to the resin, this type of damage is not of concern when GFRP rebars are used as internal reinforcement for concrete structures.

2.3.6.6 Corrosion. A major advantage of composite materials is their corrosion resistance. However, it must be noted that similar to steel, composites can also be damaged as a result of exposure to certain aggressive environments. While GFRP rebars have high resistance to acids, they can deteriorate rapidly in an alkaline environment. Concrete, particularly immediately after it is cast, is alkaline. In a recently completed study, a particular type of GFRP rebar was subjected to a saturated solution of calcium hydroxide, which resulted in high levels of water gain and loss of strength [16]. Although these results cannot be generalized, they do point to the importance of the selection of the correct fiber and matrix for the particular application. The use of alkaline resistant glass fibers is one approach to solve this problem.

2.3.7 Thermal expansion

Reinforced concrete itself is a composite material, where the rebars act as the strengthening fiber and the concrete as the matrix. It is therefore imperative that the behavior under thermal stresses for the two materials be similar so that the differential deformations of concrete and the reinforcing rebars are minimized. Depending on the mix proportions, the linear coefficient of thermal expansion for concrete varies from 6×10^{-6} to 12×10^{-6} per °C [17]. As listed in Table 2.1, the coefficient of thermal expansion for GFRP rebars is very similar to that of steel and therefore compatible for concrete construction.

2.3.8 Specific gravity

GFRP rebars have a specific gravity ranging from 1.5 to 2.0; i.e. they are nearly four times lighter than steel. The reduced weight of the materials leads to lower transportation, storage, handling on the job site, and installation compared to steel rebars. This is a major advantage of GFRP rebars which must be included in any cost analysis for material selection.

2.4 Structural behavior

Since the mid-1980s, there has been renewed interest in the application of composite materials in civil engineering. Many research projects have been undertaken in Europe, North America and Japan to investigate various aspects of concrete structures reinforced with fiber composites. The following provides a brief overview of some of these findings.

2.4.1   Flexural behavior

The behavior of concrete beams where GFRP rebars are used as flexural reinforcement is dominated by the behavior of the GFRP rebar and depends on the flexural reinforcement ratio. These concepts can be easily demonstrated with the aid of test results such as those shown in Figure 2.4.

The data presented in Figure 2.4 pertain to two beams with a rectangular cross-section of $200 \times 460\,mm^2$. The beams were simply supported on a span of 3050 mm and were subjected to two equal loads, symmetrically placed about the midspan [18]. In order to separate the shear and flexural behaviors, these beams were reinforced with adequate steel stirrups to prevent any shear failure. GFRP rebars were used only as flexural reinforcement.

The beam identified as 'under-reinforced' contained two 13-mm bars as flexural reinforcement. After the beam cracks, its stiffness is reduced and the response is linear up to the failure of the beam, caused by the fracture of the longitudinal bars. In this case, due to the relatively small tension force in the rebars at failure, the compressive strains in the concrete remain low, i.e. within the linear range. Consequently, the load–deflection response of the beam is linear. The narrow flexural cracks distributed throughout the beam indicated good bond between the GFRP rebar and concrete.

The longitudinal reinforcement for the 'over-reinforced' beam consisted of four 20-mm GFRP rebars. In this case, the large tension force developed in these bars calls for concrete compressive strains beyond the linear

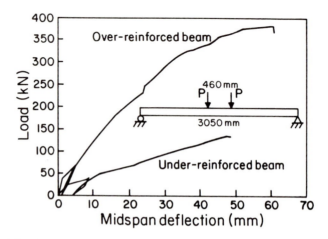

Figure 2.4 Load versus midspan deflection for beams having GFRP rebars as flexural reinforcement.

Figure 2.5 Flexural failure of an over-reinforced beam.

range. The failure of the beam is governed by the compressive strength of the concrete. Therefore, the load–deflection of such beams becomes non-linear prior to the failure, after the concrete compressive strain exceeds a value of approximately 0.002. The failed beam shown in Figure 2.5 includes a large number of cracks distributed along the entire span which indicates good bond between the GFRP rebar and concrete. Similar behavior has been observed by other researchers [10].

The flexural strength of the beams can be predicted using the principles of strain compatibility and linear variation of strain along the depth of the cross-section. For the beams reported here, there was good agreement between the measured and calculated values. Therefore, the flexural strength of beams reinforced with GFRP rebars can be calculated based on methods familiar to all engineers, taking into account the properties of the constituent materials.

2.4.2 Shear behavior

Tests have also been conducted on beams where GFRP stirrups were used as shear reinforcement. As a part of the same study discussed in Section 2.4.1, three beams were constructed where the flexural reinforcement consisted of two 22-mm steel bars [18]. Each beam was reinforced with closed rectangular stirrups made of GFRP rebars to resist the shear. A different stirrup spacing was used for each beam. The load versus midspan deflection for one of the beams, where the stirrups were placed at a uniform spacing of 150 mm is shown in Figure 2.6. The beam exhibited a large number of flexural cracks and a few flexural-shear cracks. This is

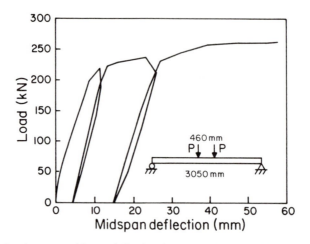

Figure 2.6 Load versus midspan deflection for beams having GFRP stirrups as shear reinforcement.

Figure 2.7 Failure of a beam having GFRP shear stirrups.

similar to the behavior of the beam if it were reinforced with steel stirrups.

As seen in Figure 2.6, the flexural steel in the beam yielded at a load of 220 kN. No major distress due to shear stresses or large shear cracks were observed. Failure of the beam occurred after the compressive strength of the concrete was exceeded. The beam after failure is shown in Figure 2.7.

2.4.3 Bond behavior

Bond characteristics of GFRP rebars to concrete are among the most important aspects of these materials. It must be noted, however, that there are no reported bond failures of GFRP rebars in field applications. Nonetheless, because of the various surface deformation patterns and textures that are present, this concern remains. Due to the lack of industry standards and the numerous proprietary manufacturing techniques that exist, a wide range of products are available on the market. Clearly these products have different bond characteristics. The subject is under investigation by several researchers and in some cases preliminary design guidelines have also been presented [10, 19–24]. In some cases, the manufacturers have developed test data to determine the bond performance of their own products. Professional organizations, e.g. the American Concrete Institute, have also established committees to develop guidelines for bond of GFRP rebars.

Some of these concerns can be best explained in the context of examining the data from an ongoing study of the bond of GFRP rebars by Ehsani et al. [22]. Figure 2.8 contains the load versus slip for a straight GFRP bar with a diameter of 19 mm and an embedment length of 300 mm. Here, the net slip, defined as the difference between the slip at the loaded end and that at the unloaded end is reported. The bar carried a load of nearly 130 kN, at which time the resin in the bar between the concrete specimen and the grips started to break and testing was stopped.

For steel reinforcing bars, allowable bond stress is defined as that associated with a slip of 0.25 mm at the loaded end or 0.05 mm at the free end [25]. Among the reasons for adoption of these criteria were the need to limit the crack widths in concrete structures to minimize the possibility

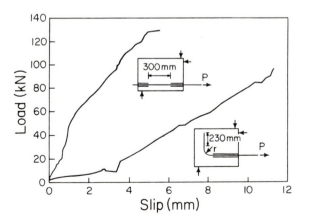

Figure 2.8 Load versus slip for bond tests of GFRP rebars.

of corrosion of steel. Clearly, for GFRP rebars, this concern is not as valid. In addition, the ribs on GFRP rebars contain a large amount of the resin, which has a low shear modulus. Thus, the contribution of the ribs in GFRP rebars to the resistance mechanism is significantly lower than that of steel rebars. Moreover, the lower modulus of elasticity of the GFRP bars contributes to a larger measured 'apparent' slip in tests. These facts point out to the need for establishing new criteria for evaluation of bond performance of GFRP rebars.

Data for a standard 90° hook are also shown in Figure 2.8. The bar has a diameter of 19 mm, a radius of curvature of $r = 57$ mm and an extension of 230 mm [22]. A comparison of the two curves indicates that the stiffness, i.e. slope, for the straight bar is larger than that of the hooked bar. The standard hook did develop a large force before the bar failed outside of the concrete block.

For steel rebars, limits have been set for the radius of curvature of the hook to prevent fracture of the bar. For GFRP rebars, where the bars are manufactured prior to the setting of the resin, this limitation does not apply. However, due to the low shear strength of GFRP rebars, if the radius, r, is small, the bar will fail in shear at the location of the hook at very small loads. This suggests that it may be advantageous to utilize even larger radii for GFRP hooks than those commonly used for steel hooks.

2.4.4 Fatigue behavior

One of the main potential applications for GFRP rebars is as reinforcement in bridge decks where the addition of de-icing chemicals has resulted in rapid corrosion of steel and deterioration of the decks. Behavior of structural components constructed with GFRP rebars and subjected to fatigue loading would clearly be of interest for such applications. As mentioned previously, a great deal of information is available on performance of composites subjected to fatigue loading. Since the fatigue response of concrete elements is greatly dependent on the fatigue life of the reinforcing material, the available data can be used to predict the fatigue life of such concrete structures.

2.4.5 Ductility

A major concern with the use of GFRP rebars may be the lack of ductility of these materials. It is noted, however, that while the rebars themselves fail in a brittle mode, the response of beams reinforced with these materials is quite different. Thus, the issue of ductility is one that requires special attention.

For concrete beams reinforced with steel, deflection ductility is defined as the ratio of the deflection at failure to that at yield of reinforcement. The term ductility relates to two aspects of the behavior of a member. The first aspect is failure under static loading at large deformations. In this regard, concrete beams reinforced with GFRP rebars do perform in a ductile manner. For the two beams shown in Figure 2.4, for example, the failure deflections were about 0.015–0.020 times the span length. Such deflections are much larger than the upper limits allowed by most codes. In fact, because of the low modulus of elasticity of GFRP rebars, the design of these structures may, in many cases, be controlled by the allowable deflections.

The second aspect of ductility is the ability to dissipate energy through inelastic deformation. This is one of the most important attributes of steel, especially in design for seismic forces where the input earthquake energy must be dissipated through inelastic deformation of the members. The elastic behavior of GFRPs prevents them from this mode of contribution. For structures expected to resist earthquake forces, a combination of GFRP rebars and external energy-dissipating devices may prove to be effective. However, at this point there are no published data on the behavior of such structural systems.

2.5 Cost

It is clear that in today's competitive global market, the survival and success of any new construction material is greatly dependent on its cost. However, initial cost of the materials cannot be used as the sole factor in making such comparisons. Other expenses associated with durability, maintenance, transportation and handling on the job site must also be included in these analyses.

Among all FRP materials available today, GFRPs are the least expensive. This is primarily due to the low cost and abundance of glass fibers. At the same time, the volume of materials needed for civil engineering applications are typically quite large. Consequently, GFRPs, whether in the form of rebars, plates, or other structural shapes offer great potential for wide applications in the construction industry.

Prices for GFRP rebars vary from one producer to the next based on the method of manufacturing, the volume and type of fibers used, etc. A price comparison based on weight or volume of materials is also meaningless because most manufacturers add a premium for special shapes which are more labor intensive such as hooks, stirrups, spirals, etc. As an example, the prices per linear foot of straight rebars for one US manufacturer ranges from $0.27 for a No. 2 (6 mm) rebar to $2.05 for a No. 9 (29 mm) rebar [8]. It is noted, however, that as discussed earlier, GFRP rebars are stronger than steel rebars. Any cost comparison should also consider the reduction in the required area of reinforcing material as a result of utilizing the stronger GFRP rebars.

After consideration is given to all advantages of these rebars, including high strength, light weight, non-conductivity, corrosion resistance and long-term savings in repair and maintenance cost, it is evident that there are many applications where the higher initial cost of GFRPs can be easily justified.

2.6   Field applications

Field applications of GFRP rebars have been consistently increasing since the mid-1980s. As more and more design engineers become familiar with the properties of these materials, they discover circumstances where GFRP rebars can be a suitable substitution for conventional reinforcement. Clearly, it is impossible to provide a list of all projects where GFRP rebars have been used. Some designers, manufacturers or contractors have documented their applications adequately while records for others are scarce at best. The intent here is to provide the reader with a sampling of the actual as well as potential applications for these materials.

The earlier applications of GFRP rebars in the United States were primarily for the construction of magnetic resonance imaging (MRI) facilities in hospitals. Steel reinforcing bars would interfere with the magnetic field of these devices and are therefore unacceptable. GFRP rebars have been used in the construction of many of these projects as shown in Figure 2.9.

More recently, there has been a surge of applications where corrosion of reinforcement is of concern. The list of users provided by a major US manufacturer includes several large chemical companies where GFRP rebars can be used for construction of floor slabs and chemical storage tanks. The rebars have been used in other cast-in-place projects such as construction of sea-walls and foundations [8].

GFRP rebars have also been used in the construction of precast concrete elements. Examples include the construction of architectural pieces in Florida where previous use of steel reinforcement had resulted in unsightly corrosion stains. In another application, a plant near Toronto makes use of GFRP rebars in construction of precast highway traffic barrier walls, where corrosion of steel reinforcement could shorten the expected life of the elements considerably.

The largest public works project undertaken to date in the United States, is the construction of a two span bridge in West Virginia. The deck of this bridge will be reinforced with GFRP rebars and the performance of the materials will be monitored for several years. Funding for this bridge has been approved and its construction will start in the second half of 1993.

Figure 2.9 Application of GFRP rebars in construction of MRI facilities (courtesy of R.D. Roll).

There are many recorded cases where grids made of GFRP have been used as reinforcement for concrete structures. The behavior of these materials is similar to that of GFRP rebars. The applications of GFRP grids, which are presented in the following chapter, also provide insight on potential uses of GFRP rebars.

2.7 Design example

In this section, sample calculations for the analysis of a typical beam are presented. The design of concrete members reinforced with GFRP rebars can be performed using the allowable stress method or the ultimate strength method. In the following example, the adequacy of the flexural design of a beam is verified according to the allowable stress method presented in Appendix A of the ACI 318-89 [26]. It is assumed that the reader is familiar with design of concrete structures according to the allowable stress method; such methods are covered in most reinforced concrete textbooks [27].

A rectangular beam with a cross-section of 300 × 600 mm² is simply supported on a span of 6 m. The beam is subjected to a total uniformly distributed service load of 23.3 kN/m, which results in a moment of 105 kN-m at the midspan. Four 19-mm GFRP rebars are placed at an effective depth of 550 mm. Concrete has a compressive strength of 40 MPa and from Table 2.2, the tensile strength of the GFRP rebars is assumed to be 620 MPa with a modulus of elasticity of 48 GPa. Verify the adequacy of the design.

With concrete having a modulus of elasticity of 30 GPa, the modular ratio, n, is calculated as 48/30 = 1.6. The total area of four GFRP rebars is A_b = 1136 mm², resulting in a reinforcement ratio of ρ = 0.00688. The factor k is then calculated as 0.138, giving a depth to the neutral axis, kd = 76 mm. The moment arm between the resultant of the compression force and tension force is $jd = d - (kd/3)$ = 525 mm.

Next, the stresses in the reinforcement, f_b, and concrete, f_c, can be calculated and compared with the allowable values:

$$f_b = \frac{M}{(A_b)(jd)} = \frac{105 \times 10^6}{(1136)(525)} = 176 \text{ MPa}$$

$$f_c = \frac{(2)(M)}{(b)(kd)(jd)} = \frac{(2)(105 \times 10^6)}{(300)(76)(525)} = 17.5 \text{ MPa}$$

The allowable compressive stress is 45% of the concrete compressive strength or 18 MPa. For the GFRP rebar, the allowable tensile stress is conservatively assumed to be 35% of the tensile strength of the bar or 217 MPa. The calculated stresses for concrete and GFRP rebar are both within the allowable limits.

A check is also performed on the deflection of the beam. The gross moment of inertia for the beam is I_g = 540 000 cm⁴. The cracked moment of inertia, I_{cr}, can be calculated as

$$I_{cr} = \frac{b(kd)^3}{3} + nA_b(d - kd)^2$$

$$I_{cr} = \frac{(30)(7.6)^3}{3} + (1.6)(11.36)(55 - 7.6)^2 = 45\,230 \text{ cm}^4$$

With the modulus of rupture, f_r, for the concrete taken as 3.8 MPa, the cracking moment for the beam, M_{cr}, can be calculated as

$$M_{cr} = \frac{(f_r)(I_g)}{y_t} = \frac{(3.8 \times 10^3)(540\,000)}{30} \times \frac{1}{10^6} = 68.4\,\text{kN-m}$$

For midspan deflection due to the dead and full live load, the effective moment of inertia, I_e, is

$$I_e = \left[\frac{68.4}{105}\right]^3 (540\,000) + \left[1 - \left(\frac{68.4}{105}\right)^3\right](45\,230) = 182\,000\,\text{cm}^4$$

This maximum deflection in the simply supported beam due to the uniformly distributed load is equal to

$$\Delta = \frac{5(23.3)(6.0)^4}{(384)(30 \times 10^6)(182\,000 \times 10^{-8})} = 0.0072\,\text{m} = 7.2\,\text{mm}$$

This deflection is nearly 1/800 times the span and is well within acceptable limits.

2.8 Summary and conclusions

An overview of GFRP rebars has been presented in this chapter. This includes the manufacturing and mechanical properties of the bar itself, as well as the behavior of structural members reinforced with GFRP rebars. While a great deal of information has been accumulated in recent years about these materials, much more research needs to be completed before the advantages and shortcomings of GFRP rebars are fully understood. It is therefore advisable that, in the meantime, adequate attention be given to the selection of the factors of safety for various aspects of design.

There is no doubt that composites are rapidly becoming the next generation of construction materials. The applications demonstrated here attest to the wide range of potential possibilities for the use of GFRP rebars. It is incumbent upon the design community to become familiar with the advantages and limitations of these materials and to foster the growth of this field through its prudent use.

References

1. Gustafson, D.P., Epoxy update, *Civil Eng.* **October** (1988) 38–41.
2. Keesler, R.J. and Powers, R.G., Corrosion of epoxy coated rebars—Keys Segmental Bridge—Monroe County, Report No. 88-8A, Florida Department of Transportation, Materials Office, Corrosion Research Laboratory, Gainsville, 1988.
3. Crepps, R.B., Glass fibers as the tensioning element in prestressed concrete. *Proc 1st U.S.A. Conf. on Prestressed Concrete*, 1951, pp. 228–230.
4. Rubinsky, I.A. and Rubinsky, A., A preliminary investigation of the use of fiberglass for prestressed concrete. *Mag. Concrete Res.* **6**(17) (1954) 71–78.

5. Somes, N.F., Resin bonded glass fiber tendons for prestressed concrete. *Mag. Concrete Res.* **15**(45) (1963) 151–158.
6. Nawy, E.G., Neuwerth, G.E. and Phillips, C.J., Behavior of fiberglass reinforced concrete beams. *J. Struct. Div.* ASCE **September** (1971) 2203–2215.
7. Mufti, A.A., Erki, M.A. and Jaeger, L.G., eds., Canadian Society of Civil Engineers, 1991, 297 pp.
8. Fiberglass reinforced plastic rebar data sheet, KODIAK, International Grating Inc., Houston, Texas, 1992.
9. GangaRao, H.V.S. and Barbero, E., Construction, structural applications, in S.M. Lee, ed., *International Encyclopedia of Composites*, Vol. 6, VCH, New York, 1992.
10. GangaRao, H.V.S. and Faza, S.S., Bending and bond behavior and design of concrete beams reinforced with fiber reinforced plastic rebars, Final report to the Federal Highway Administration, West Virginia University, 1991, 159 pp.
11. Wu, W.P., Thermomechanical properties of fiber reinforced plastic bars, Ph.D. Dissertation, West Virginia University, 1991.
12. Bedard, C., Composite reinforcing bars: assessing their use in construction. *Concrete Int.* **January** (1992) 55–59.
13. Budelmann, H. and Rostasy, F.S., Creep rupture behavior of FRP elements for prestressed concrete—phenomenon, results and forecast models, *ACI Int Symp. on FRP Reinforcement for Concrete Structures*, Vancouver, Canada, 1993.
14. Schwartz, M.M., *Composite Materials Handbook*, McGraw-Hill, New York, 1992.
15. Franke, L., Behaviour and design of high-quality glass-fiber composite rods as reinforcement for prestressed concrete members, Report, International Symposium, CP/Ricem/i Bk, Prague, 1981, pp. 171–174.
16. Zayed, A.M., Deterioration assessment of fiber-glass plastic rebars in different environments, *The NACE Annual Conf. and Corrosion Show*, Cincinnati, OH, 1991, pp. 130/1–130/8.
17. Mehta, P.K., *Concrete: Structure, Properties, and Materials*, Prentice Hall, Englewood Cliffs, NJ, 1986, p. 88.
18. Saadatmanesh, H. and Ehsani, M.R., Fiber composite bar for reinforced concrete construction. *J. Composite Mater.* **25** (1991) 188–203.
19. Pleimann, L.G., Strength, modulus of elasticity, and bond of deformed FRP rods, in S.L. Iyer, ed., *Advanced Composite Materials in Civil Engineering Structures*, ASCE, 1991, pp. 99–110.
20. Chaallal, O. and Benmokrane, B., Pullout and bond of glass-fiber rods embedded in concretes and cement grout. *RILEM* **26**(157) (1993).
21. Daniali, S., Development length for fiber-reinforced plastic bars, in K.W. Neale and P. Labossiere, ed., *Advanced Composite Materials in Bridges and Structures*, Canadian Society of Civil Engineering, 1992.
22. Ehsani, M.R., Saadatmanesh, H. and Tao S., Bond of GFRP rebars to ordinary-strength concrete, *ACI Int. Symp. on FRP Reinforcement for Concrete Structures*, Vancouver, Canada, 1993.
23. Nagel, F., Untersuchungen zum Trag- und Verformungsverhalten von mit HLV-Elementen vorgespannten Stahlbetonzugkörpern, M.Sc. Thesis, IWB, Universität Stuttgart, 1986, 187 pp. (in German).
24. Mashima, M. and Iwamoto, K., Bond characteristics of FRP rod and concrete after freezing and thawing deterioration, *ACI Int. Symp. on FRP Reinforcement for Concrete Structures*, Vancouver, Canada, 1993.
25. Park, R. and Paulay, T., *Reinforced Concrete Structures*, Wiley, New York, 1975, pp. 407–409.
26. ACI Committee 318, Building code requirements for reinforced concrete (ACI 318-89), American Concrete Institute, Detroit, Michigan, 1989, 353 pp.
27. MacGregor, J.G., *Reinforced Concrete*, Prentice Hall, Englewood Cliffs, NJ, 1988, pp. 304–334.

3 NEFMAC grid type reinforcement

M. SUGITA

3.1 Introduction

New fibre composite material for reinforced concrete (NEFMAC) is an innovative concrete reinforcement consisting of high strength continuous fibres such as carbon fibre, glass fibre and aramid fibre. All of them are impregnated with resin and formed into two- or three-dimensional grid shapes.

NEFMAC is very lightweight, stronger than steel reinforcement, does not rust or corrode and has a very high resistance to salt. The characteristics of NEFMAC are shown in Table 3.1. In addition, Figure 3.1 shows an example of the form of NEFMAC and Figure 3.2 shows a close up of the intersection of a grid.

3.2 Manufacturing process

NEFMAC is formed into flat or curved two- or three-dimensional grid shapes by the newly developed pin-winding process, which is a kind of filament winding process [1]. In a batch process, to form large cross-sectional or three-dimensional grid shapes, fibres are impregnated with resin activated by a peroxide curing system, and formed into the grid shape at room temperature in successive layers as shown in Figure 3.3. In this process, the resin takes a few hours to cure. In the case of a continuous process to form a two-dimensional grid with a small cross-section, fibres are impregnated with a resin incorporating an ultraviolet curing system, and formed into a flat grid as shown in Figure 3.4. Since the line speed is 0.5–1.5 m/min, 60–180 m²/h of NEFMAC can be produced by this method.

3.3 Survey of development

Table 3.2 is a survey of NEFMAC research and development (R&D) up to the present [2]. Work was begun in 1984. At first, basic R&D was done independently and then joint R&D was done with universities to clarify NEFMAC's fundamental properties. Later, independent R&D and joint

Table 3.1 Characteristics of NEFMAC

• Non-corrosive	• Improvement of the durability of concrete structures used under severe conditions where damage by salt and chemicals is to be expected
• Excellent resistance to alkalis, acids, and chemicals	
• Usage of continuous fibres	• Effective usage of fibres giving a hybrid effect[a] by mixtures of different kinds of fibres
• Assuring the strength of the intersections of the grid	• Assuring sufficient anchorage to the concrete
	• Availability of lapped splice joint
• Light weight (specific gravity $\doteq$ 2)	• Improving productivity in the field
• Possibility of forming intricate shaped objects in a single piece	
• Non-magnetic	• Applicability to a structure requiring non-magnetic properties

[a] Hybrid effect: in a composite material which is reinforced with more than two kinds of fibres, break starts from the fibre with the least elongation and other breaks take place in order. The final break in the material occurs when the fibre with the largest elongation breaks. In this process, the stress–strain relationship becomes non-linear and shows the same phenomenon as the yielding of a reinforcing steel bar.

R&D with universities was carried out to apply NEFMAC to actual concrete structures.

As for fundamental R&D, we clarified NEFMAC's tensile properties and conducted studies that related to the spliced lap joint mechanisms [3]. These results proved that NEFMAC is effective as reinforcement for concrete structures.

In the joint R&D which was carried out with universities, shear and compressive properties, i.e. NEFMAC's fundamental mechanical characteristics, were clarified. The mechanism of anchoring to the concrete was then determined. It became clear that the shear strength is about 50% of the tensile strength in the case of Type H [3], combined glass and carbon fibres, and that anchorage capacity greater than the tensile strength can be achieved by NEFMAC provided two pieces of transverse reinforcement are embedded into concrete [3], and so on. Additionally, tests on chemical resistance under constant tensile deformation [1], tests on conduction in a spa atmosphere [6], bending tests after sustained loading and so on were carried out which confirmed NEFMAC's durability. Moreover, to define the behaviour of concrete reinforced with NEFMAC in fire and at high temperature, NEFMAC's resistance to heat was also studied.

As the next stage of our technological development programme, we carried out tests in collaboration with universities to verify the application

Figure 3.1 An example of the form of NEFMAC.

Figure 3.2 Intersections of a grid.

of NEFMAC to concrete structures. For wall panel development, in-plane bending-shear tests under reversed cyclic lateral loading and bending tests under out-of-plane loading were carried out, and structural characteristics clarified. For slab panel development, beam-anchoring methods, flexural behaviour and long-term deflection were confirmed by testing. In addition, to improve the flexural behaviour, we studied the possibility of prestressing by pretensioning the reinforcement [4]. The fire resistance of wall and slab panels was studied and the suitability of NEFMAC in this application was verified. For applications to columns and beams, the flexural, shear and fatigue behaviour and especially flexural behaviour after long-term exposure were confirmed.

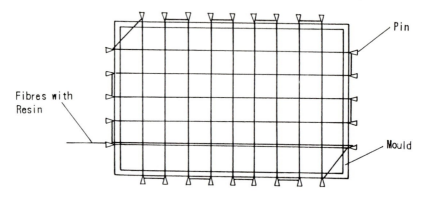

Figure 3.3 Batch process for the production of NEFMAC.

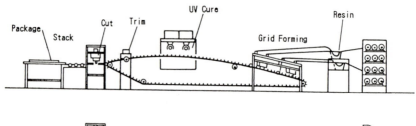

Figure 3.4 Continuous process for the production of NEFMAC.

3.4 Material properties of NEFMAC

3.4.1 Tensile and compressive properties

The standard specification of NEFMAC is shown in Table 3.3. In this table, 'max. load' is the guaranteed value. Further information is given in [5]. Examples of tensile test results of NEFMAC are shown in Figure 3.5.

The tensile testing methods and the compressive testing method for NEFMAC are shown in Figures 3.6 and 3.7 respectively. For tensile testing, three grip types are adopted. With any grip type, the tensile test goes well and the tensile strength of NEFMAC is evaluated correctly. For compressive testing, the method of confining the buckling of NEFMAC by concrete blocks is adopted.

The relation between the data for the tensile and compressive maximum

Table 3.2 Research and development items for NEFMAC

Purpose	Examination
Clarification of fundamental properties of NEFMAC	
Mechanical properties	Tensile strength of fibres
	Tensile strength and hybrid effect
	Compressive strength
	Shear strength
	Reversed cyclic loading (tension/ compression)
	Strength of cross point
Bond mechanism	Anchorage to concrete
	Lap splice joint
Durability	Chemical resistance under constant tensile deformation
	Tensile strength under spa atmosphere
	Creep fracture
Heat resistance	Tensile strength under/after heating/cooling
Proof of the applicability to concrete structures	
Earthquake resisting wall	Behaviour under reversed cyclic lateral loading
Precast wall structure	Shear resistance of vertical joint
Wall panel	Flexural behaviour
	Fire resistance
Slab panel	NEFMAC's anchorage to beam
	Time-dependent deflection
	Flexural behaviour
	Fire resistance
	Effectiveness of prestressing
Beam and column	Effectiveness of main and shear reinforcement
	Confined effect by lateral reinforcement
	Fatigue behaviour
	Flexural behaviour after sustained loading
	Flexural behaviour with exposed NEFMAC in air
	Effectiveness of chemical prestressing
Shotcrete and concrete lining	Work productivity

loads is shown in Figure 3.8. The relation between the data for the tensile and compressive stiffness is shown in Figure 3.9. The symbols represent the average values and the arrows represent the ranges of the data for the five specimens. From these figures, it is clear that the following conclusions may be drawn. In the case of type G NEFMAC (glass fibre), the data for the compressive maximum load are almost equal to the data for the tensile maximum load. In the case of type C NEFMAC (carbon fibre), however, the compressive data are about 40% of the tensile ones.

Table 3.3 Standard specification of NEFMAC

Type	Bar no.	Sectional area (mm^2)	Max. load (tonf)	Tensile strength (kgf/mm^2)	Young's modulus (kgf/mm^2)	Weight (g/m)
G	G2	4.4	0.26			7.5
	G3	8.7	0.52			15
Glass fibre	G4	13.1	0.78			22
+ resin	G6	35.0	2.1	60	3000	60
	G10	78.7	4.7			130
	G13	131	7.8			220
	G16	201	12.0			342
	G19	297	17.7			510
H	H6	39.5	2.1			65
	H10	88.8	4.7			147
Glass fibre/	H13	148.0	7.8	53	3700	244
carbon fibre	H16	223	12.0			368
+ resin	H19	335	17.7			553
	H22	444	23.4			733
C	C6	17.5	2.1			25
	C10	39.2	4.7			56
Carbon fibre	C13	65.0	7.8	120	10000	92
+ resin	C16	100	12.0			142
	C19	148	17.7			210
	C22	195	23.4			277
A	A6	16.2	2.1			21
	A10	36.2	4.7			46
Aramid fibre	A13	60.0	7.8	130	5700	77
+ resin	A16	92.3	12.0			118
	A19	136	17.7			174

As for the stiffness, the compressive and tensile data are almost equal regardless of the type of NEFMAC.

3.4.2   Durability of NEFMAC

3.4.2.1   Durability test conducted in a spa atmosphere

Purpose of the test.   The durability of NEFMAC had been confirmed in tests based on the JIS K 7107 Test Methods for Chemical Resistance of Plastics Under Constant Tensile Deformation, for which various chemicals were used including calcium hydroxide and sodium chloride. Its actual in-use durability, however, had not been proved. With the co-operation of the Ohita Construction office of the Japan Highway Public Corporation, durability tests on NEFMAC were performed at the Beppu Alum

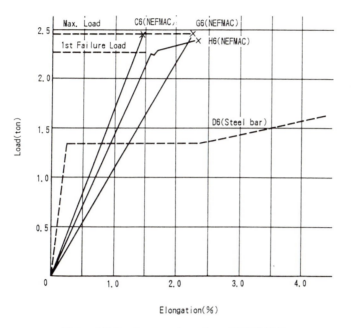

Figure 3.5 Load versus elongation of NEFMAC.

Spa [6]. The remainder of this section is based on the data obtained from the first to the twelfth month of the tests.

Test method. Types and quantities of the test pieces are given in Table 3.4. Each test piece was installed so that 300 mm remained buried underground and 600 mm were left exposed to the air (surface environment). Nameplates were attached to each group of the test pieces for the purpose of classification. Installed test pieces are shown in Figure 3.10. Visual examination of the specimens was carried out at intervals. The corroded state was observed and recorded and photographs were taken. To determine the tensile strength, three bars were collected from each sample by cutting away the lateral bars. These bars were cut and separated into buried and surface sections. Samples were subjected to a tension test at Matsuzaki Laboratory, Architectural Division, Faculty of Engineering, Science University of Tokyo. Maximum load and fracture conditions were recorded. NEFMAC test pieces were reinforced at the gripped sections and then tested.

Test results.

1. Underground temperature distribution and pH: the temperature distribution below the surface of the ground and the results of pH

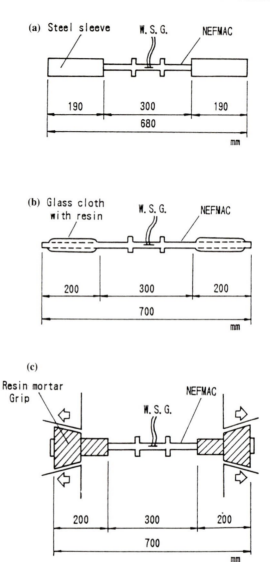

Figure 3.6 Tensile testing method of NEFMAC: (a) steel-sleeve grip type; (b) glass-cloth grip type; (c) resin-mortar grip type.

measurements at the place where test pieces were buried are shown in Figure 3.11. Temperature was maintained at about 80°C at the lowest horizontal bar, roughly 60°C at the middle horizontal bar and about 50°C at the horizontal bar near the surface. The maximum tensile load capacity of the buried test pieces was taken as the strength of a section heated to approximately 60°C.

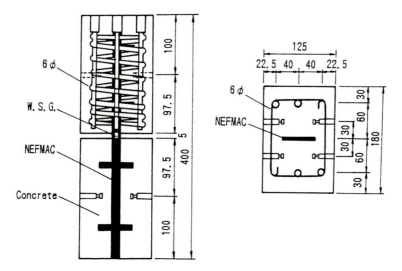

Figure 3.7 Compressive testing method of NEFMAC.

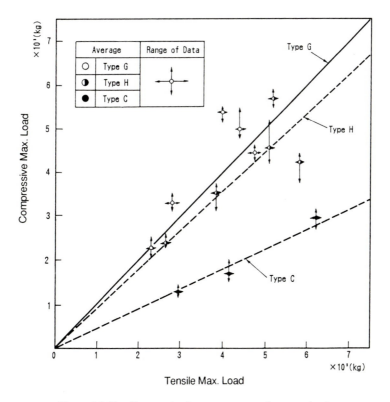

Figure 3.8 Tensile max. load versus compressive max. load.

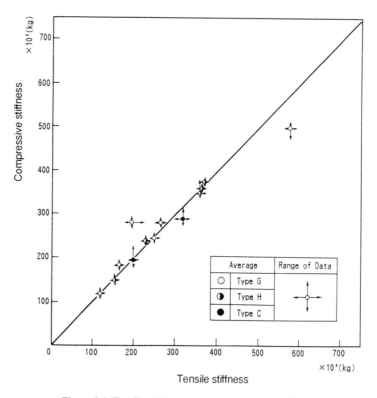

Figure 3.9 Tensile stiffness versus compressive stiffness

Table 3.4 Types and quantities of exposed test pieces

Diameter (mm)	Material[a] (pieces)		
	NEFMAC	Welded wire mesh	Steel reinforcement
4	11 (G4)	11 (4φ)	–
10	11 (G10)	–	11 (D10)

[a] Each material was tested to confirm initial strength.

2. Results of visual examination: as can be seen in Table 3.5, there were almost no changes in the surface sections of the NEFMAC samples. However, changes were seen in the buried underground sections, among them a progressive increase in colouring that resulted from being buried deeper. This led to blackening of the lowest horizontal bar. The colour change could hardly be noticed after 2 months; thus, the degree of colouring was considered to be

Figure 3.10 Installed test pieces.

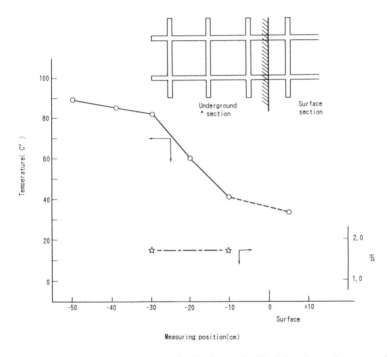

Figure 3.11 Underground temperature distribution and pH of the place where test pieces were buried.

governed by the underground temperature. The surface sections of the welded wire mesh and steel reinforcement samples had rust on them; after 2 months, the state of the rust did not change. However, temperature and pH greatly influenced the buried sec-

Table 3.5 Results of visual examination

Test piece		Initial state March 3, 1988	2 months later May 18, 1988	4 months later July 6, 1988	6 months later September 21, 1988	12 months later March 4, 1989
NEFMAC G4	Surface	Translucent camel colour	No change	Same as left	Same as left	Same as left
	Under-ground	Same as above	Lowest horizontal bar was blackened	No change in colouring	Same as left	Same as left
NEFMAC G10	Surface	Same as above	No change	Same as left	Same as left	Same as left
	Under-ground	Same as above	Lowest horizontal bar was blackened	No change in colouring	Same as left	Same as left
Welded wire mesh 4φ	Surface	Rust-free	Rust (red) generated	No change in rusted state	Same as left	Same as left
	Under-ground	Same as above	Partially melted away	Dissolved portion increased	Mostly dissolved away	–
Steel-reinforcement D10	Surface	Same as above	Rust (red) generated	No change in rusted state	Same as left	Same as left
	Under-ground	Same as above	Section reduced by rusting	Section further reduced by rusting	Partially dissolved away	Mostly dissolved away

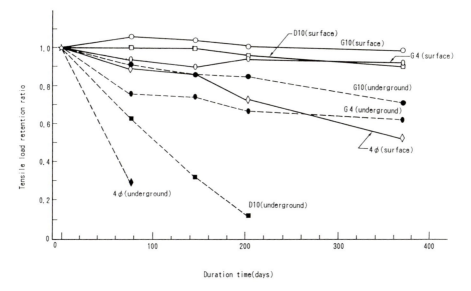

Figure 3.12 Tensile load retention ratio and duration.

tions. Dissolving of the test piece could be noticed 4 months later for welded wire mesh and 12 months later for steel reinforcement. Thus, tension tests could not be run on those samples.

3. Results of tension test: from the tensile load retention ratio (Figure 3.12), it is clear that the surface section of NEFMAC G4 and G10 and that of steel reinforcement D10 do not lose strength even after 12 months in the atmosphere of the exposure site. In contrast, the tensile load retention of welded wire mesh 4ϕ was reduced to roughly 50% because of rust. When buried sections came into contact with soil of about 60°C and a pH of 1.7 (Figure 3.11), corrosion and deterioration was more striking than in the above ground sections. For welded wire mesh and steel reinforcement, corrosion progressed in a linear way. The 4ϕ welded wire mesh dissolved away in about 100 days and D10 steel reinforcement in about 220 days. With NEFMAC, strength reduction could be seen after 12 months while G10 held onto 70% for the same period. Since G10 has a large area, more time is required to reach the 60% plateau of the retention ratio of G4.

3.4.2.2 *Behaviour of beams subjected to sustained load* [7]

Test methods.

1. Beam specimens and loading method: The dimension of the beams and the loading method are given in Figure 3.13. Reinforcement

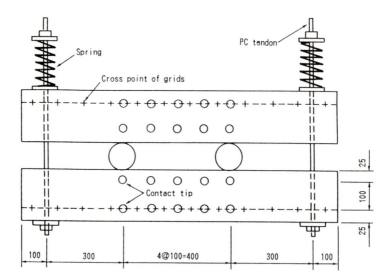

Figure 3.13 Test method of sustained loading.

was arranged in only the axial direction. The distance from the tension fibre to the reinforcement was 25 mm.

2. Types of NEFMAC and mechanical properties: three types of NEFMAC, type C, type G and type H were used (carbon fibre, glass fibre and combined glass and carbon, respectively). They were formed into a grid shape with a 10 cm pitch. The criteria for the strength of the NEFMAC were always 1.2 times the tensile strength of reinforcing steel (D10, SD345).

Test results.

1. Deformation behaviour under sustained load: the strains of the reinforcement and the average curvatures under sustained load were smaller the greater the tensile stiffness of the reinforcement. Also a clear difference was seen between the rates of change in curvature and in the deformation behaviour of steel reinforced and NEFMAC beams subjected to sustained loads. In Figure 3.14 the change in curvature after loading is shown. The absolute values of curvature, as mentioned previously, were higher with the beams using NEFMAC of low tensile rigidity than with the RC beams. This is thought to have been because the flexural cracking of NEFMAC beams developed rapidly to reach close to the compression fibre so that the increase in curvature accompanying development of cracks while subjected to subsequent sustained load was small.

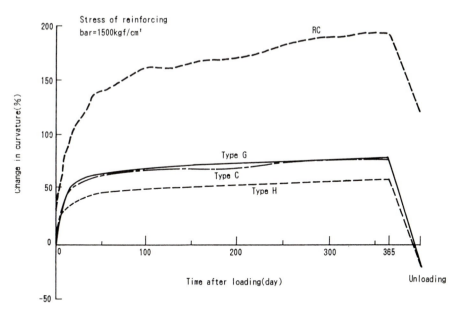

Figure 3.14 Change in curvature.

2. Mechanical behaviour of beams subjected to different sustained loads: flexural strength tests were performed on beams after loads had been sustained for 1 year. Strains in the reinforcement, flexural crack widths, deflections and failure loads were determined. The relationships between average flexural crack widths and loads for type G (glass fibre) beams subjected to different sustained loads are shown in Figure 3.15. The crack width at a load of 0 tf corresponds to the residual crack width immediately before the flexural strength test was carried out. The arrows in the figure indicate values of initial loads during sustained load tests. The increases in flexural crack width for beams under large sustained loads, since flexural cracking during sustained loading had developed considerably, showed roughly the same trends, and the effects of different initial loads could hardly be seen. Beams subjected to a sustained load equal to a stress of $1000\,kgf/cm^2$ in the reinforcing steel in equivalent RC beams, had larger width flexural cracks at low load when compared with beams tested at 28 days of age without being subjected to sustained loads and beams with load of 0 tf applied, and they became closer to beams with heavy sustained loads. As a result of the above, it was recognised that with NEFMAC beams, the flexural behaviour in flexural strength tests after sustained loading was influenced by whether flexural cracks had developed

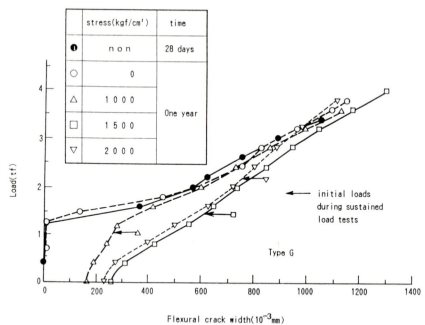

Figure 3.15 Relationship between flexural crack width and load.

considerably or not. There were cases where they were hardly affected by the magnitude of the sustained load.

3.5   Structural characteristics of NEFMAC reinforced concrete

3.5.1   Flexural behaviour

To apply NEFMAC reinforced concrete panels to curtain walls and fender plates for small ships, we conducted bending tests to clarify the flexural behaviour [2]. Table 3.6 shows the physical properties of NEFMAC and the concrete used in testing.

To develop curtain walls, we confirmed the wind resistance of the precast concrete panels by bending tests. The details of the specimen are shown in Figure 3.16 and load–deflection relationship of specimens in Figure 3.17. Bending tests were also done on specimens which had undergone fire resistance testing in accordance with the Japanese Industrial Standard; Figure 3.17 shows the test results.

The aforementioned tests clarified that the flexural cracking load exceeds the design load and that the maximum strength of the precast

Table 3.6 Material properties of bending tests

Applications	Materials	Type	Max. (yield) load (ton)	Stiffness (ton)
Curtain wall	NEFMAC	H8	3.79	224
		G8	3.10	155
	Lightweight concrete	$F_c = 303\,\text{kgf/cm}^2$	–	–
Fender plate	NEFMAC	G19	17.6	886
	Reinforcing bar	D19 (SD295A)	Yield = 10.3	5707
	Concrete	$F_c = 161\,\text{kgf/cm}^2$	–	–

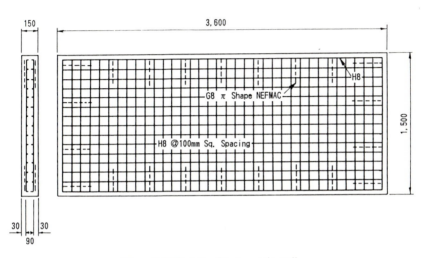

Figure 3.16 Details of test curtain wall.

concrete panels is only slightly reduced after fire resistance testing. We also learned that the existing bending theory for reinforced concrete can be applied to calculate the initial stiffness (I_g), post-cracking stiffness (I_{cr}) and maximum strength (P_u). Additionally, we clarified through both experiments and analysis that the allowable crack width under the design load is 0.3 mm or less, even if there is flexural cracking.

Fender plates for small ships were originally designed with epoxy-coated steel reinforcing bars. However, the comparison of flexural behaviour (exhibited in bending tests) of two precast concrete panel specimens, one with epoxy-coated steel reinforcing bars and the other with NEFMAC reinforcement, led us to use NEFMAC in some of the panels. The dimensions of the specimens are shown in Figure 3.18, physical properties of the materials used in Table 3.6 and the load–deflection relationship of the specimen in Figure 3.19. As NEFMAC has

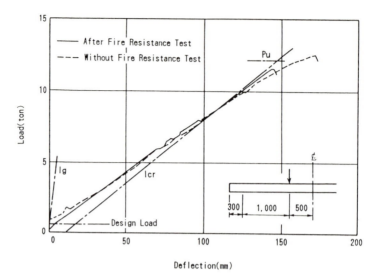

Figure 3.17 Load–deflection relationships of test curtain walls.

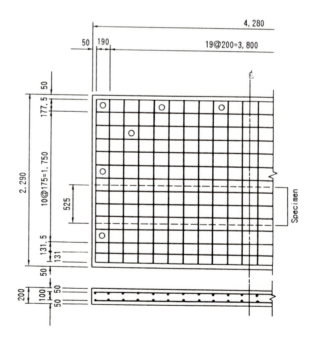

Figure 3.18 Reinforcement of fender plate.

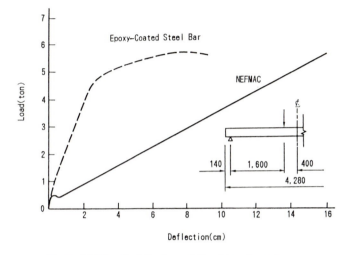

Figure 3.19 Load–deflection relationships of specimens.

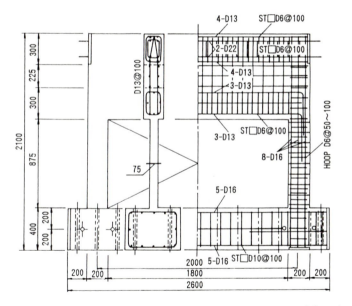

Figure 3.20 The dimensions of the specimens and the bar arrangement of the columns and beams.

a small Young's modulus, the bending stiffness of the specimen dropped after flexural cracking. The deflection was approximately five times larger than for the epoxy-coated steel reinforcing bars while they were below the yield (i.e. within the elastic range). Thus, we understood from this

Table 3.7 Material properties of shear walls

Materials	Type	Max. (yield) load (ton)	Stiffness (ton)
NEFMAC	H6	2.52	152
	H3	1.23	78
Reinforcing bar	D16	Yield = 7.20	3611
	D6 (column)	Yield = 1.30	594
	D6 (wall)	Yield = 1.22	599
Concrete	$F_c = 212 \sim 233\,\mathrm{kgf/cm^2}$	–	–

result that the NEFMAC reinforced specimen has a large capacity for absorbing energy, an advantage for a fender plate function.

3.5.2 Shear behaviour

Trials have been carried out on three shear walls [8]. The specimens were $2.1 \times 2.6\,\mathrm{m^2}$ overall, with a thin central panel, as shown in Figure 3.20. The edge beams and columns were reinforced with conventional steel, as detailed in Figure 3.20 and Table 3.7. Three different reinforcement arrangements were used for the wall panels, one using steel and the other two using NEFMAC Type H (glass and fibre combined), as shown in Figures 3.21 to 3.23. The loading arrangement, shown in Figure 3.24, was such that both vertical and shear loads could be applied to the specimen.

The test results are given in Figures 3.25–3.27, which show that the behaviour of all three specimens was similar, with the exception of the maximum deformation. These tests showed that NEFMAC should be suitable for the reinforcement of thin shear walls.

3.6 Applications

As of the end of March 1992, 1 150 000 m² of NEFMAC had been used for concrete reinforcement. The following is a discussion of its major uses. Further information is given in [9, 10].

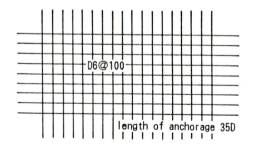

Figure 3.21 The bar arrangement in the wall (No. 1).

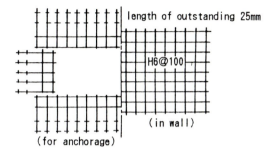

Figure 3.22 The bar arrangement in the wall (No. 2).

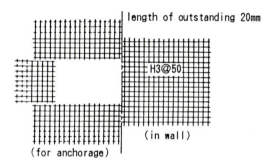

Figure 3.23 The bar arrangement in the wall (No. 3).

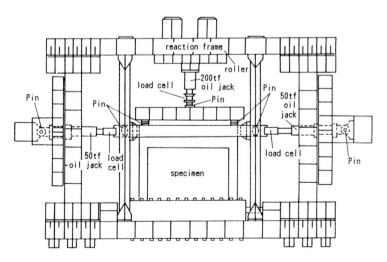

Figure 3.24 Loading method.

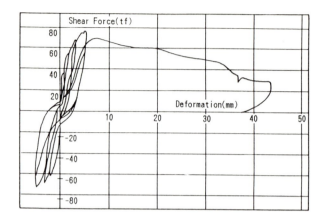

Figure 3.25 The relations between load and deformation (No. 1).

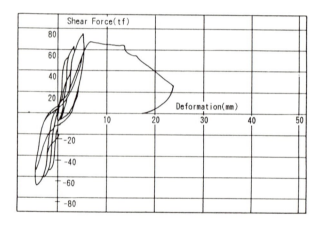

Figure 3.26 The relations between load and deformation (No. 2).

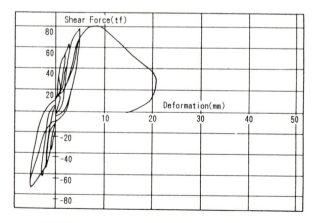

Figure 3.27 The relations between load and deformation (No. 3).

Figure 3.28 Oil storage rock tanks.

Figure 3.29 Railway tunnel.

Figure 3.30 Invert of water conveyance tunnel.

Welded mesh is sometimes used to reinforce shotcrete in the New Austrian Tunnelling Method (NATM). However, due to its high rigidity, large gaps are caused between the welded mesh and the primary shotcrete. Therefore, more secondary shotcrete is needed to fill in the large gaps. This increases costs. Additionally, a lot of labour is required to fix the mesh in position. In contrast, due to NEFMAC's low rigidity, only small gaps are caused, so less secondary shotcrete is required which helps cut cost. The simplified method that has been developed to set NEFMAC by air gun stapling has significantly reduced the installation time and overall construction period [10]. Because of its high durability, NEFMAC has been used for reinforcing shotcrete for oil storage tanks in rock (Figure 3.28). NEFMAC has also been used to reinforce shotcrete in the repair works on a railway tunnel (Figure 3.29). Its non-conductivity was recognised and proved especially useful in the prevention of electric shock.

In general, the shotcrete of the tunnels is lined with concrete as with road tunnels. This lining is often reinforced with steel bars; however, the tunnel's narrowness demands a significant amount of time and labour for arranging and fixing reinforcement bars. However, NEFMAC, being lightweight and formed into either curved or flat planes in factories, drastically reduces work inside tunnels. Figures 3.30 and 3.31 are examples of NEFMAC applications; arrangement time was reduced by approximately 70%.

Because NEFMAC does not rust, it is a promising concrete reinforcement material for ocean coastal structures. Figure 3.32 shows a concrete pontoon. A block of styrene foam is contained in the concrete skin. Even if cracks are formed in the concrete skin, the pontoon does not sink. To allow for cracking of the concrete, rust free reinforcement is adopted.

Figure 3.31 Arch of water conveyance tunnel.

Figure 3.32 Pontoon.

Figure 3.33 Cushion board.

Figure 3.34 Control building in the Antarctic Base of Japan.

Figure 3.35 OA floor.

Figure 3.36 Foundation of Earth Magnetism Observatory.

Figure 3.33 depicts the application of NEFMAC for the concrete cushion panels for small vessels, which were attached to a landing stage. NEFMAC has also been applied to the stop logs for the water intake of a steam power plant. Two years have passed since NEFMAC was used for reinforcing the pontoon, 3 for the reinforcing panels and 6 years for reinforcing the stop log; no problems have occurred so far.

Figure 3.34 shows the construction site of the control building in the Antarctic base of Japan. This building had to be constructed by the members of the expedition within a short period of time; they were not trained for construction. NEFMAC was adopted because of its prefabrication and light weight.

NEFMAC has often been adopted because it is not magnetic and does

Figure 3.37 Curtain wall.

not conduct electricity. One example of such usage is in the reinforcement of a GRC OA floor. Approximately $160\,000\,\text{m}^2$ of NEFMAC has been used for this purpose (Figure 3.35). The major reason for its use is that NEFMAC completely inhibits the flow of stray currents and can greatly raise the amount of energy absorbed during bending of GRC. NEFMAC has also been used to reinforce the concrete used in the foundation of an earth magnetism observatory, for which no metals could be used (Figure 3.36). This kind of performance would also prove beneficial in concrete girder reinforcement in linear motor cars.

A further use for NEFMAC has been the reinforcement of a light-weight concrete curtain wall building which was approved by Japan's Minister of Construction (Figure 3.37). Although there are many legal barriers that make it difficult to adopt advanced composite materials in buildings in Japan, it is expected that the need for lightweight, highly durable building materials will increase.

References

1. Hayashi, K., Sekine, K., Sekijima, K. and Nakatsuji, T., Application of FRP grid reinforcement for concrete and soil, *46th Annual Conf. Composite Institute*, The Society of Plastics Industry, 1991, ss. 12-D, pp. 1–7.
2. Sugita, M., Nakatsuji, T., Sekijima, K. and Fujisaki, T., Application of FRP grid reinforcement to precast concrete panel, *Advanced Composite Materials in Bridges and Structures, ACMBS-1*, 1992, pp. 331–340.

3. Fujisaki, T., Sekijima, K., Matsuzaki, Y. and Okumura, H., New material for reinforced concrete in place of reinforced steel bar, *IABSE Symp.*, Paris-Versailles, 1987, pp. 413–418.
4. Sekijima, K. and Hiraga, H., Fiber reinforced plastics grid reinforcement for concrete structures, *IABSE Symp.*, Brussels, 1990, pp. 593–598.
5. Nakatsuji, T. Mechanical Properties of NEFMAC, NEFCOM Corporation Technical Leaflet, 1990.
6. NEFCOM Corporation, NEFMAC Durability test conducted in spa atmosphere, 1989.
7. Tsuji, Y., Sekijima, K., Nakajima, N. and Saito, H., Mechanical behaviors of concrete beams reinforced with grid shaped FRP and effects of chemical prestress, *Concrete Library JSCE* No. 18, 1991, pp. 211–221.
8. Fujisaki, T., Kokusho, S., Kobayashi, K., Hayashi, S., Ito, C. and Arai, M., Application of new fiber reinforced composite material (NFM) to concrete shear wall, Report of the Research Laboratory of Engineering Materials, Tokyo Institute of Technology, No. 15, 1990, pp. 529–534.
9. Nakatsuji, T., Sugita, M., Fujimori, T., FRP grid reinforcement for concrete and soil, Transportation Research Board 69th Annual Meeting, Paper No. 89 CP057, January 1990.
10. Ikeda, K., Sekijima, K. and Okamura H., New materials for tunnel supports, *IABSE 13th Congress*, Helsinki, 1988, pp. 27–32.

4   Oriented polymer grid reinforcement
G.R. CARTER

4.1   Introduction

Oriented polymer grids are lightweight, non-corrodible and effective in controlling cracking in concrete. They can therefore provide a durable alternative to steel reinforcement in non-structural applications in both aggressive and non-aggressive environments. This chapter introduces the properties and method of production of oriented polymer grids, reviews the research that has been carried out into the potential use of oriented polymer grids in concrete and describes some practical applications.

4.2   Properties of oriented polymer grids

Tensar* oriented polymer geogrids were developed during the late 1970s. Since then their use in a wide range of civil engineering applications has grown rapidly. The high strength and long-term durability of the grids has led to their extensive use in soil reinforcement, ground stabilisation and road pavement applications. In addition, the grids have been used in a range of concrete applications.

The principal function of the grids in concrete is to distribute stresses and control cracking in exposed concrete surfaces such as bonded repairs to existing structures, e.g. bridge beams, sea walls, buildings, sewer linings, etc. and as a distribution mesh in paved areas or screeds.

Tensar geogrids are manufactured from either polypropylene or high density polyethylene, two of the most stable and durable polymers. The grids are inert and have no solvents at ambient temperature. They are therefore particularly suitable for use as a reinforcement in aggressive environments.

4.3   Method of production

During the manufacturing process, sheets of polymer are punched with a regular array of holes (Figure 4.1). The sheet is then stretched under controlled temperature and strain rate conditions. During this process, the long chain molecules, which are randomly orientated within the

---

*Tensar is the registered trade mark of Netlon Limited in the United Kingdom and other countries.

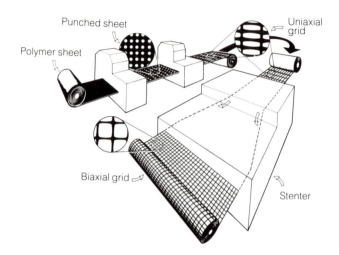

Figure 4.1 Tensar geogrid manufacturing process.

original sheet, are aligned along the direction of stretch. This results in an increase in the tensile strength and stiffness to produce uniaxial grids which are primarily used in reinforced soil applications.

If the elongated product is now stretched in the transverse direction, a biaxial grid with rectangular or square apertures is produced. These grids have an increased tensile strength and stiffness in both directions and are generally used for ground stabilisation and pavement applications and also for secondary reinforcement of concrete.

An example of an oriented polymer grid, the Tensar CR1 geogrid, is shown in Figure 4.2. It is available in 3.5 m wide rolls, 50 m long. It is lightweight and completely corrosion resistant. The high strength ribs connected by strong integral junctions achieve bond with the concrete by means of mechanical interlock through the grid apertures. This allows the grid to distribute stresses and provides cost effective crack control in many areas of concrete construction and repair.

Where polymer grids are used, the amount of cover can be significantly reduced compared with that normally required to protect steel reinforcement. By installing the grids closer to the point from which cracking develops, a more efficient means of crack control can be achieved.

4.4   Research into concrete applications

Prior to commercial scale developments within the concrete field, a 3.5-year programme of research, funded jointly by the Science and Engineering Research Council and Netlon Limited, was undertaken by

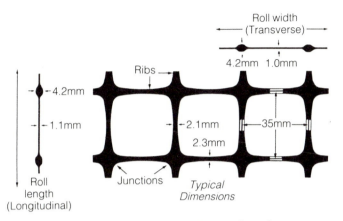

Figure 4.2 Tensar CR1 geogrid typical dimensions.

the Department of Civil and Structural Engineering at the University of Sheffield. The research team at Sheffield investigated the behaviour of concrete incorporating polymer grids using various methods of applying stress to composite panels under cyclic, static and dynamic loading conditions [1, 2]. The remainder of this part of the chapter summarises this work.

4.4.1 Static loading

Static loading involved the study of the behaviour of oriented grid reinforced concrete under compression, tension and flexure. The effects of thermal cycling were also observed.

4.4.1.1 Compression behaviour. Compressive strength tests carried out on mortar cylinders indicated increasing ductility at failure as the layers of oriented grid reinforcement increased. The ultimate strain for a cylinder reinforced with one layer of grid was more than 70 times that of a plain concrete cylinder. The energy absorption capacity was similarly increased by about 600 times for one layer of grid and 1500 times for three layers of grid compared with the plain concrete. Post-cracking load capacity was also demonstrated.

4.4.1.2 Behaviour in tension. The oriented grid reinforced specimens demonstrated both increasing load capacity after the first crack and a large capability for energy absorption.

4.4.1.3 Behaviour in flexure. The behaviour of polymer grid reinforced concrete under flexure was observed using simply supported concrete

Table 4.1 Results of flexural tests (after Swamy et al. [1])

Specimen	Plain	Steel mesh	Tensar SS1		
			1 layer	4 layers	8 layers
Load to first crack (kN)	0.64	0.63	0.61	0.52	0.49
Load to 20 mm deflection (kN)	–	–	0.50	0.76	1.11
Gross stress at first crack (N/mm^2)	3.20	3.15	3.05	2.60	2.45
Gross stress at 20 mm deflection (N/mm^2)	–	–	3.50	3.80	5.55

slabs $700 \times 300 \times 20 \, \text{mm}^3$ upon which loads were applied at third points. Test specimens incorporated an increasing number of oriented polymer grid layers, in this case Tensar SS1. For purposes of comparison, testing was also carried out on plain and reinforced specimens using 1 mm diameter steel mesh structures with a 50 mm grid pitch.

As the number of grid layers increased, cracking occurred at progressively lower loads probably because the grids are less stiff than the cement matrix and increasing layers of grid decrease the effective area of the matrix. Each specimen was taken beyond the first crack stage up to 20 mm of deflection. The plain and steel reinforced specimens collapsed before this stage, failing therefore to exhibit any post-crack ductility. The polymer grid composites however, demonstrated significant post-crack ductility, a feature which was evident even where only one polymer grid layer was used (Table 4.1).

An important feature brought out by this work is the ability of the polymer grids to distribute stresses and dissipate strains over a larger area. This contrasts with the plain and steel reinforced members which exhibited large cracks at mid-span and offered no residual flexural strength. Whilst cracks occurred more extensively as the number of grid layers increased, the spacing became noticeably finer, thus demonstrating that crack widths can be controlled by the geometry and the number of the grids.

The grid reinforced composites when reloaded after the first loading, retained a significant proportion of the first loading behaviour. They also showed almost total recovery after repeated loading and extensive cracking. This led Swamy et al. to suggest that not only would oriented polymer grid reinforced elements recover after extensive cracking and deformation, but that they would also retain their shape and integrity in a structural system. With steel mesh reinforcement, however, there was little recovery after the first loading, particularly if the steel had yielded, and further loading continued to produce plastic behaviour until failure.

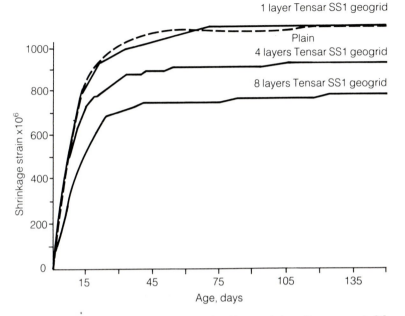

Figure 4.3 Free drying shrinkage (reproduced with permission, Swamy *et al.* [1] and Thomas Telford Publications).

Again, as with the tension and compression tests, the capacity for energy absorption increases with increasing volume of grid. Similar observations on crack spacing, load capacity after significant deflection and increasing capacity with increasing volume of grid were made in work recently carried out by Clarke [3].

4.4.1.4 Shrinkage behaviour. The free drying shrinkage of a grid reinforced composite decreases as the volume of grid increases. The results of shrinkage tests carried out on $700 \times 300 \times 20\,mm^3$ specimens are shown in Figure 4.3. A comparison with a plain unreinforced specimen is also shown. The shrinkage stabilises early in life and remains at a reasonably constant value, unless the ambient environment changes.

4.4.1.5 Thermal cycling. Plain and oriented polymer grid reinforced specimens were subjected to thermal cycling to model the effects to which many concrete units such as thin cladding panels are subjected. The results, as shown in Figure 4.4, indicated that the grid reinforced units showed both a more uniform and reduced deformational behaviour compared to plain concrete. These effects increased as the number of grid layers increased.

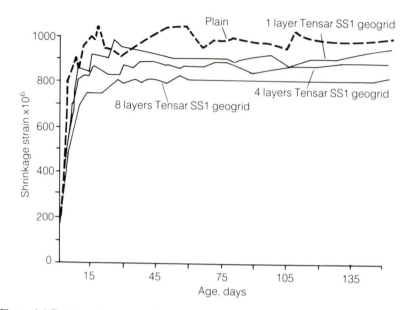

Figure 4.4 Results of thermal cycling (reproduced with permission, Swamy *et al.* [1] and Thomas Telford Publications).

The cracks appeared progressively earlier with increasing layers of grid probably due to the different coefficients of thermal expansion of the grids and the matrix and also the reduction in effective area of matrix due to the grids, but the cracks tended to close once the thermal effects had ceased.

4.4.1.6 Flexural creep behaviour. The results of flexural creep tests indicated that polymer grid reinforced composites are likely to undergo large deformations if subjected to sustained loadings over a period of time. This behaviour is to be expected in a material where the matrix is stiffer than the reinforcement. However, the tests also indicated that grid reinforced composites should be able to carry short-term loads without undue deformation. The ability of the composites to return to their original state is also unlikely to be affected by creep in the short-term.

4.4.2 Dynamic loading

The dynamic testing work carried out at Sheffield involved the application of impulsive loading from explosive charges, high velocity, low mass impacts from bullets and low velocity, high mass impacts from drop hammers. In addition to observing the transient responses and permanent damage to the concrete members, some work was carried out on static re-

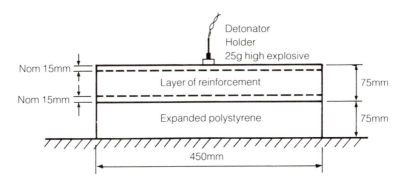

Figure 4.5 Explosive impulse experiment (reproduced with permission, Watson *et al.* [2] and Thomas Telford Publications).

loadings of the damaged slabs. This work was carried out to simulate the effects of damage due to accidental impact, mishandling or vandalism on semi-structural units which are lightly loaded under normal service conditions.

4.4.2.1 Explosive impulse. High explosive (25 g) was placed at the centre of thick concrete slabs (450 mm², 75 mm thick) as shown in Figure 4.5. An unreinforced control slab was fractured and the parts propelled some distance, but slabs containing either steel mesh or polymer grids held together, even though they were cratered and cracked (Figure 4.6).

Using normal weight concrete the crater volume on the top of the slab was marginally less and the spall volume on the underside of the slab was much less in slabs doubly reinforced with polymer grids than with steel, despite the fact that the steel can provide a greater tensile resistance. The grids were most effective in controlling crater and spall dimensions when placed close to the outer faces of the slab.

4.4.2.2 Bullet impact. In this experiment, 7.62 mm bullets were fired at the centre of a concrete slab (450 mm², 125 mm thick). Measurements of crater depth and diameter indicated that, because of the nature of the stresses imposed by the bullet, the type of reinforcement did not make a significant difference.

4.4.2.3 Drop hammer impact. In these tests a falling weight produced a point impact at the mid-spans of 700 × 300 × 20 mm³ slabs spanning 600 mm (Figure 4.7). The unreinforced slabs suffered a single crack near the mid-span on impact and collapsed totally. In the case of steel reinforcement, a crack developed across the middle of the slab as the steel yielded and the slab collapsed as the steel fractured (Figure 4.8).

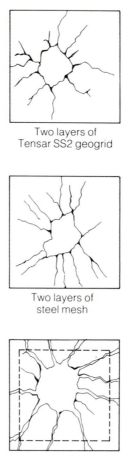

Two layers of
Tensar SS2 geogrid

Two layers of
steel mesh

Plain

Figure 4.6 Typical specimens after impulsive loading produced by 25 g high explosive (reproduced with permission, Watson *et al.* [2] and Thomas Telford Publications).

The polymer grid reinforced slabs exhibited a finer and more extensive crack pattern (Figure 4.8) and showed a greater tensile resistance as none of these slabs collapsed.

In contrast to the steel reinforced slabs, which did not oscillate but collapsed on impact, the polymer grid reinforced slabs oscillated and then damped out to leave almost no permanent deflection. The grid reinforced slabs showed a pseudo-elastic response to the single impact. Increasing layers of grid both increased the stiffness and decreased the total crack length in the slab.

Static flexural tests were subsequently carried out on the damaged specimens and by comparing their performance with that of previously

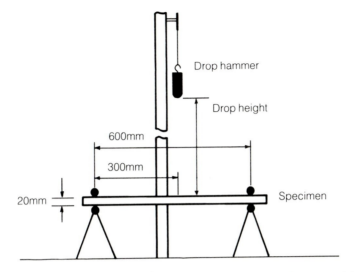

Figure 4.7 Drop hammer test apparatus (reproduced with permission, Watson *et al.* [2] and Thomas Telford Publications).

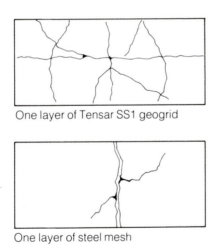

Figure 4.8 Impacted face of panels after drop hammer test (reproduced with permission, Watson *et al.* [2] and Thomas Telford Publications).

untested members, residual strengths varying between 40% and 90% were recorded, depending upon the volume fraction of the grid. This again demonstrates the characteristic of post-crack ductility in concrete incorporating polymer grids.

Further work on impact testing of steel and polymer grid reinforced slabs has been carried out at Carleton University in Canada by Suter *et al.* [4]. In this research programme, steel blocks weighing 17 kg and 34 kg were repeatedly dropped on the centre of simply supported concrete slabs at ever increasing heights until the slabs failed by breaking into two or more pieces. The slabs were reinforced either by polymer grid, steel wire fabric or a combination of the two. The position of the reinforcing layers was also varied.

The results of the tests indicated that the polymer grid reinforced slabs were superior to the steel reinforced specimens in terms of both impact resistance and resilience. Initial deflections of the polymer grid reinforced slabs were normally greater than those reinforced with steel. This means that the polymer grid reinforced slabs have a greater capacity to absorb energy. The residual or permanent deflection for the polymer reinforced specimens was lower than for the steel reinforced slabs. As also shown by the Sheffield work, the deformation in the polymer reinforced slabs was mostly elastic up to failure.

Another interesting observation which came out of the Carleton study was that when the polymer grid reinforcement was combined with steel reinforcement in a single slab, the polymer grids were found to be more effective in enhancing the capacity of the slabs to endure blows when placed in the tension zone at the bottom of the slab.

4.4.3 Structural performance of oriented polymer grid reinforcement

The static and dynamic testing programme carried out at Sheffield led to a number of conclusions being drawn about the properties of polymer grid reinforced concrete composites.

In tension, flexure and compression, the polymer grid reinforced concrete composites were found to possess considerable post-cracking load capacity, extensive ductility and substantial load carrying capacity. Additionally, the cracked slabs were able to recover almost completely on unloading and retain their integrity, and in some cases, retain a substantial proportion of their load carrying capacity.

The crack spacing was dominated by the spacing of the ribs of the outermost grid, suggesting that a systematic and uniform crack spacing could be obtained by suitably sized grids.

The presence of the polymer grids reduced shrinkage movements and tended to stabilise such movements earlier than unreinforced specimens. Under thermal cycling, the polymer grid reinforced composites showed reduced deformational behaviour and although the grids caused cracking to occur earlier, once the thermal movements ceased, the composites were able to return to their apparently uncracked state.

The major conclusion that has been drawn as a result of the research is that in situations other than those where sustained loads have to be carried over a long period of time, oriented polymer geogrids offer a suitable alternative to steel reinforcement.

4.5 Practical applications of oriented polymer geogrids

As a result of the potential benefits of oriented polymer geogrid reinforcement identified by the research, a number of practical application areas were identified. These were mainly in the field of repair to existing structures, particularly in aggressive environments, where the grid is used as a non-corrodible carrier for either mortar or sprayed concrete. However, applications in new works have also been carried out, particularly in the area of ground supported slabs. The remainder of the chapter consists of a number of short case studies describing the use of oriented polymer grids in actual projects.

4.5.1 Marine situations

4.5.1.1 Sea wall repair. The ability of oriented polymer grids to conform to sweeping curves and unusual shapes is of particular benefit in the application of sprayed concrete. This property was used to good effect on a contract at Walton-on-the-Naze, Essex, where oriented polymer geogrids were specified by Tendring District Council in the repair of a sea wall suffering from extensive spalling due to reinforcement corrosion [5]. A 75-mm thick layer of gunite was applied by the wet process after installation of the grid. Correct location of the grid was achieved by the use of standard fixings which allow a predetermined stand off from the underlying concrete. The gunite therefore completely envelopes the grid, the resulting composite providing a non-corrodible protective coating to the sea wall (Figure 4.9). The repair is behaving satisfactorily more than 6 years after installation.

4.5.1.2 Repairs to piles. Concrete members exposed to continuous wetting and drying cycles or occasional impact forces from marine craft can suffer serious corrosion problems, often necessitating urgent remedial attention. At Ilfracombe, the piles supporting part of the pier were repaired by the use of fabric jackets filled with a superplasticised 'micro concrete' [6]. Two layers of Tensar oriented polymer grid were placed within the 100–150 mm thick annulus to control cracking in the concrete encasement. The grids were separated from each other and the existing structure by means of spacer blocks (Figure 4.10).

Figure 4.9 Sprayed concrete repairs to sea wall, Walton-on-the-Naze.

Figure 4.10 Two Tensar grid layers are used with a fabric former, Ilfracombe pier.

4.5.2   External insulation of property

Extensive use has been made of oriented polymer geogrids in the external insulation of buildings. Not only does external insulation allow properties to be improved with minimal disturbance to the occupants, it also dramatically improves the thermal efficiency of all external walls. It is a method particularly suited to the upgrading of reinforced concrete system built dwellings.

If a property is externally insulated the position of the dew point is within the external render. This means that interstitial condensation will form within the render which will effectively become a warm damp medium. The carrier mesh supporting the render must therefore be non-corrodible. Stainless steel mesh is sometimes used in this application. However, the use of an oriented polymer grid in this situation can provide a more cost effective and, because of its lighter weight and larger roll size, a more convenient solution. The cost of the polymer grid reinforcement is approximately half that of the stainless steel.

Various methods of fixing can be employed to attach the grid to the substrate and the insulant. One such fixing device, in this case made of polypropylene with a stainless steel core, is shown in Figure 4.11.

Both sprayed or hand-applied mortar can be used. Figure 4.12 shows the applications of a silicate enriched mortar mix devised by Michael Dyson Associates for the external refurbishment of system built houses for Walsall Metropolitan Borough Council [7]. After the spray concrete has cured, a pebble dash finish is applied.

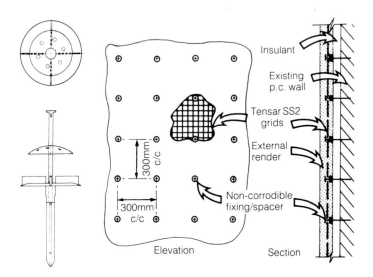

Figure 4.11 Detail of grid fixings in external insulation of buildings.

Figure 4.12 Applying spray concrete to an oriented polymer grid carrier on Walsall Metropolitan Borough Council properties.

Figure 4.13 A waterproof hand-applied base coat covers the carrier grid (acknowledgement Epsicon Ltd).

Epsicon Ltd have developed a system whereby a 15–20 mm thick render is hand applied in two coats. A waterproof base coat covers the carrier grid (Figure 4.13). This is then scratched for bond prior to applying the top coat and finishings. The in-service behaviour of this type of insulating system is reported to be good.

4.5.3   Repair of structures

Oriented polymer geogrids have also been used in the repair of highway structures. An example of this application is the repair of three crossheads supporting an elevated carriageway on the Coventry Inner Ring Road. This work was carried out in 1985 [8]. Here the concrete had deteriorated due to chloride corrosion caused by exposure to de-icing salts over a number of years. The concrete was beginning to spall, which could have led to structural weakening.

The repair method consisted of the removal of 100 mm of damaged concrete. This was then replaced with 95 mm of sprayed concrete. This silicate enriched mix, which was designed by Michael Dyson Associates, provided a waterproof layer which was intended not only to prevent the passage of additional chloride ions into the concrete but also to immobilise the ions already present.

Work on these repairs was restricted to alternate 1.2-m wide sections of the crosshead at any one time. The appearance of the surface of this 95-mm thick layer of sprayed concrete was therefore non-uniform. A further 25-mm final covering of silicate enriched mortar was then sprayed onto the entire crosshead, thus increasing the depth of cover of the structural reinforcement by 20–120 mm. This outer covering was reinforced with a Tensar oriented polymer grid (Figures 4.14, 4.15). The non-corrodible nature of the polymer grid allowed the cover to be reduced to a minimum. The use of this reinforcement in the outer layer reduced the possibility of large cracks forming, which would allow further ingress of aggressive liquids. This repair is performing satisfactorily some 7 years after installation.

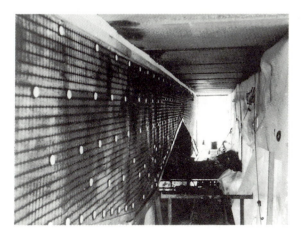

Figure 4.14 Tensar geogrid fixed to crosshead.

Figure 4.15 Application of sprayed concrete outer layer.

4.5.4   Use of oriented polymer grids in tunnels

4.5.4.1   Sewer rehabilitation.   Oriented polymer grids have been used in sewer rehabilitation work both in sprayed and precast lining systems [9]. As the grids can be placed very close to the surface of the concrete, the concrete thickness can be reduced substantially whilst still providing a durable lining (Figures 4.16, 4.17). This is of particular benefit in the stabilisation or strengthening of old brick sewers where it is important that the reduction in flow capacity is minimised.

In addition, the high energy absorption capacity of the polymer grid reinforced concrete is of benefit in reducing damage caused by waterborne debris. For this reason, Tensar grids have been used, cast close to the face, to augment galvanised steel mesh in precast concrete culvert invert units which may be subjected to high current velocities.

4.5.4.2   New construction.   The wedge segmental method of construction employed by Transmanche Link for the Channel Tunnel running tunnels leaves a small key space of approximate dimensions $1 \times 0.2 \times 0.29\,\text{m}^3$ in the roof of the tunnel at each key location. This gap is infilled with grout during the grouting of the cavity behind the tunnel segments using a shutter placed flush with the segments. In order to prevent this grout fragmenting and causing pieces to fall from the roof, some reinforcement is required to hold the grout together. The use of metallic reinforcement was ruled out due to the possibility of corrosion.

Tensar polymer grid reinforcement was selected both for its non-corrodible properties, and for its ability to be cut and bent easily to shape. Individual pieces of grid were supplied pre-cut and were bent *in*

Figure 4.16 Tensar grids are installed in a brick-lined sewer prior to spraying using a wet gunite process.

Figure 4.17 Application of sprayed concrete to a Tensar carrier grid in a brick culvert.

situ to form the shape of the actual key (Figure 4.18). The pieces of grid were fixed 25 mm from the shutter face by means of lightweight spacers.

4.5.5 Ground supported slabs

The ability of oriented polymer grids to control cracking allow them to be used as a distribution mesh in a variety of paving and flooring applications such as industrial floors, hardstandings and marine slipways.

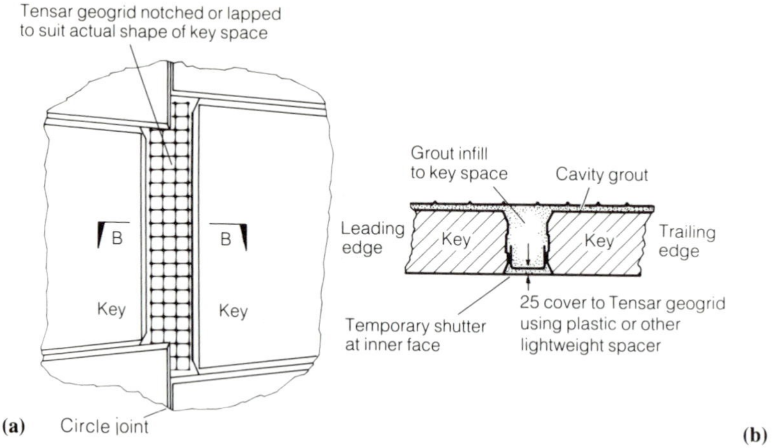

Figure 4.18 Detail of grid fixing in grouted key space within Channel Tunnel running tunnels: (a) view on key space in tunnel roof; (b) section B–B (acknowledgement Transmanche Link).

In marine or exposed conditions, it is common to increase cover depths simply to protect metallic reinforcement from corrosion and therefore prevent surface disruption of the concrete. Using oriented polymer grids, it is possible to place the reinforcement closer to the surface, and therefore closer to the potential cracking zone. The relatively close spacing of the ribs of the grid will produce finer cracks than steel mesh reinforcement. These cracks would not normally be wide enough to conduct moisture. If, however, moisture ingress did occur, the polymer grids would be unaffected and there would therefore be no reinforcement induced disruption of the concrete.

Two examples of the use of polymer oriented grids in ground supported slabs in maritime conditions are a yacht club slipway and causeway at Erith [10] and part of the promenade at Swansea.

Oriented polymer grids are also of benefit in some types of industrial flooring where the use of steel reinforcement would cause electro-magnetic disturbance of either computer installations or remote controlled vehicles. An example of this situation was at the Asda Stores regional distribution centre at Dartford. Guide cables were to be set into the cold store floor slab for the control of self-driven stacking vehicles. No steel reinforcement was therefore allowed within 80 mm of the surface of the concrete. Approximately 3500 m^2 of Tensar geogrid was placed 30 mm from the surface of the running slab, which was itself constructed on top of insulation separating it from the structural floor beneath.

4.6 Conclusions

Since the development of oriented polymer grids over a decade ago, a significant amount of research has been carried out to discover their potential for the reinforcement of concrete. The research identified that the grids could be of benefit for crack control purposes in a number of applications where substantial loads are not required to be carried over a long period of time. Such applications include concrete repair, ground supported slabs and the reinforcement of thin non-load-bearing panels. The non-corrodible nature of the grids makes them particularly suitable for use in aggressive environments.

Oriented polymer grids have been used in a number of these applications, most particularly in the field of concrete repair. One area, however, which has not yet been significantly developed is in the use of polymer grids to augment steel reinforcement in load-bearing concrete units, and particularly those which may be required to take impact loading. The research has shown that the polymer grids can significantly increase the resilience, ductility, impact resistance and energy absorption capabilities of conventional reinforced concrete.

References

1. Swamy, R.N., Jones, R. and Oldroyd, P.N., The behaviour of Tensar reinforced cement composites under static loads, *Symposium on Polymer Grid Reinforcement in Civil Engineering*, ICE, London, 1984, pp. 222–232.
2. Watson, A.J., Hobbs, B. and Oldroyd, P.N., The behaviour of Tensar reinforced cement composites under dynamic loads, *Symposium on Polymer Grid Reinforcement in Civil Engineering*, ICE, London, 1984, pp. 233–242.
3. Clarke, J.L., Tests on slabs with non-ferrous reinforcement, FIP Notes, 1992/1, pp. 5–7 and 1992/2, pp. 2–4.
4. Suter, G.T., Abd El-Halim, A.O., Winterink, J.A. and Rahman, A.H., Comparative impact testing of steel and polymer geogrid reinforced concrete, *Proc. Int. Conf. on Structural Faults and Repair*, University of London, 1987, pp. 365–372.
5. Tensar grids in concrete repairs in the marine environment, *Concrete* **21**(8) (1987) 31.
6. Ingram, C.J., *Repair of Marine Structures. Maritime and Offshore Structure Maintenance*, Thomas Telford, London, 1986, pp. 113–126.
7. External refurbishment and thermal upgrading of system built dwellings at Heather Road, Bloxwich, Netlon Limited Case Study, 1985.
8. Middleboe, S., Ringroad trials may herald £4M repairs, *New Civil Eng.* **21 February** (1985) 4–5.
9. Morris, M., Some of the applications for polymer grids in concrete, *Concrete* **21** (2) (1987) 25–26.
10. Construction of a concrete slipway and causeway at Erith Yacht Club, Kent, Netlon Limited Case Study, 1984.

5 Parafil ropes for prestressing tendons

C.J. BURGOYNE

5.1 Introduction

Parafil ropes have several features which distinguish them from most other prestressing systems; they cannot be bonded to concrete; they contain no resin, and they were not initially developed for prestressing. Nevertheless, they have been used for prestressing concrete on a number of occasions, and have recently been adopted by one of the largest manufacturers of prestressing systems (VSL) as an alternative to steel tendons when corrosion is likely to be a problem.

Unbonded tendons are already widely used in floor slabs in buildings, but have not been widely used in bridge construction in the past. This has been due to worries about corrosion of the steel. However, the UK Department of Transport has recently imposed a moratorium on the use of grouted tendons in bridges, because of concern that grouting is rarely fully effective, leading to potential for steel corrosion in the voids. They propose that tendons should be unbonded, which although it makes the steel more susceptible to corrosion, allows them to be removed and re-placed. Parafil, however, offers the same inspectability and replaceability, but in an inherently non-corroding material. There are also technical reasons why materials that do not yield should be unbonded; these are discussed later.

5.2 Description

Parafil ropes are manufactured by Linear Composites Ltd in Yorkshire, England. They contain a core of parallel filaments of a high strength yarn within a polymeric sheath. A variety of core yarns are used, the most common being polyester (known as Type A), Kevlar 29 (Type F) and Kevlar 49 (Type G). Kevlar was the first of the aramid fibres to be developed, by EI DuPont de Nemours, in 1973. The basic properties of ropes manufactured with these yarns are shown in Table 5.1. Those of primary interest to prestressing engineers are the Type G ropes, which have the highest stiffness and lowest creep properties, although the lower modulus versions could be used in applications where the prestressed structure itself tended to creep under the influence of the prestress. In this case, the lower modulus of the fibres would require larger jack ex-

Table 5.1 Tensile properties of Parafil ropes (data from Linear Composites Ltd)

Designation	Material	Strength (N/mm^2)	Stiffness (kN/mm^2)
Parafil Type A	Polyester	617	12.0
Parafil Type F	Kevlar 29	1926	77.7
Parafil Type G	Kevlar 49	1926	126.5

tensions at the time of prestressing, but would mean less loss of prestress due to creep.

Other fibres could also be used, including the alternative aramid fibres such as Technora (made by Teijin), or possibly Vectran (made by Hoechst). Both claim to have certain properties better than Kevlar, and there is no reason why they could not be used in these ropes.

5.3 Termination system

The most important component of any system carrying tension is the anchorage, where the forces are transmitted to the rope. In Parafil, the ropes are anchored by means of a barrel and spike fitting, which grips the fibres in an annulus between a central tapered spike and an external matching barrel (Figure 5.1).

To attach the termination, the end of the rope is passed through the terminal body, and the sheath is removed over the length of the spike; the yarns are then spread out evenly around the terminal body before the spike is introduced. The rope is drawn back into the terminal and the rope pretensioned to a load in excess of that to be applied in practice. During pretensioning, the spike is drawn fully into the termination, applying an outward force on the fibre pad and gripping the yarns. Subsequent changes in force cause only tiny movements in the spike and can be ignored for practical purposes. For prestressing operations, where the largest load applied to the rope is the act of prestressing, the pretensioning of the rope can take place at the same time as the prestress. Allowance then has to be made for the bed-down of the spike to ensure that the rope is the correct length, but this is easily done.

This system has a number of advantages over wedge systems which grip the outside of a tension member.

1. The gripping force between the spike and the barrel has to pass through every fibre (Figure 5.2), which means that each fibre can develop an equal friction force against its neighbours or the fitting. Thus, there is no tendency for some of the fibres to carry a dis-

Figure 5.1 Barrel and spike termination for Parafil.

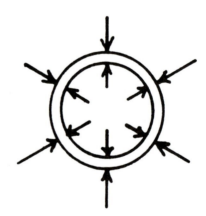

Figure 5.2 Gripping forces within termination.

proportionate amount of the load, which would cause early failure of those fibres, and hence the rope. Systems which rely on external wedges have a tendency to develop hoop compression around the outside of the tension member, leaving the inner fibres less well gripped.

2. There is no resin in the system, which means that the effectiveness of the termination is not affected by temperature or creep.
3. The system is easy to fit, on site if necessary, simply by removing the sheath and splaying out the fibres. If possible, a pretension in excess of that expected in the service life of the ropes should be applied.
4. There are no size effects; terminations for large ropes are linearly scaled versions of the terminations for small ropes. The mechanics of operation remain the same.
5. The terminations can develop the full strength of the parent rope; when used for tension tests, the rope breaks away from the termination.

There appears to be no degradation of termination efficiency, or of creep within the termination, with time.

5.4 Development of ropes

The ropes were first developed in the early 1960s to meet a requirement for mooring navigation platforms in the North Atlantic. These would have required mooring lines several kilometres long and the weight of steel ropes would have been prohibitive. At the same time, accuracy of position of these platforms meant that the lines had to be stiff and to have good axial fatigue performance. Conventional structured ropes, where the individual fibres follow tortuous paths along the rope, could not be used since they lose a significant proportion of the fibre's inherent stiffness. The large number of points where fibres cross also causes a loss of fatigue strength.

Linear Composites Ltd (then part of Imperial Chemical Industries) developed the idea of keeping the fibres straight and giving the rope some structure (normally provided by braiding or twisting) by enclosing the fibres in an extruded sheath. The fibres used at that time were polyester, the aramids not yet being available.

In the event, the requirement for aircraft navigation systems was met by satellites, so the North Atlantic platforms were never needed, but methods of producing the ropes and their properties were well established, and this led to their adoption in a variety of applications. The earliest of these were as guys for radio antennae where the non-conducting nature of the ropes did away with the need for conventional insulators. The first installations used polyester ropes, but since the development of aramid fibres, and with the communications industry using arrays of masts which have to be placed accurately in relation to one another, the stiffer aramid ropes are now being used more extensively.

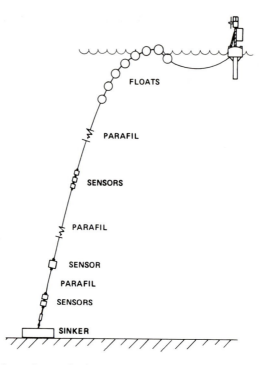

Figure 5.3 Typical mooring application, in 6000 m water depth, showing Parafil combined with floats and other equipment.

Moorings for floating systems, such as buoys, are used extensively (Figure 5.3). These often have floats, weights, and other equipment attached at the top and bottom, with the bulk of the length of the mooring being provided by Parafil.

Other early uses of Parafil as replacements for steel wire followed. Such uses have included standing rigging in ships, where the smooth sheath has been found to provide the added advantage that ice can easily be shaken free, supports for overhead wires in trolley-bus systems (again making use of the electrical insulation), and in safety rails around the deck of ships. Many of these systems are in use by military authorities around the world.

In the 1970s, Kevlar, the first of the aramid fibres, became available. Experiments showed that the techniques used for making ropes from polyester could easily be adopted for aramids, resulting in a stronger and stiffer rope. The strength of the rope is about 20% higher than a normal prestressing steel, while the stiffness is about two-thirds that of steel. These properties make the ropes very attractive as structural elements in their own right, and a programme of research was undertaken to give practising engineers confidence in both the short- and long-term properties of the ropes.

Prestressing tendons for concrete were soon identified as a very suitable application. These tendons are the most heavily stressed elements in normal use; no other structural component is regularly loaded to a permanent force of 70% of its break load. For this use, there are clear advantages in using the stiffer Type G Parafil, incorporating Kevlar 49, and in all that follows, it is this type of Parafil that is being considered, unless otherwise stated.

5.5 Testing

Much of the early testing on Kevlar concentrated on the short-term properties of the fibre, but for structural engineering applications, the long-term properties are just as important. A lot of testing has thus been done on the ropes rather than the fibres, to establish their properties.

5.5.1 Strength and size effects

The stress−strain curve of the ropes (Figure 5.4) matches quite closely that of the constituent fibres. The Young's modulus is about $120 \, kN/mm^2$ and the strength is about $1930 \, N/mm^2$. There is a slight stiffening at about $1000 \, N/mm^2$; that is a property of the fibre and is not significant in most cases. Once the rope has been fitted with terminals and these have been bedded down properly in accordance with the manufacturer's instructions, the ropes have their full stiffness from zero load.

When a number of fibres are used together, it is not possible to use the full strength of all the fibres, or even to achieve the average strength of the fibres. This is because the weaker fibres fail at a lower load than the stronger ones, leaving the total load-carrying capacity reduced. This process has to be applied twice in Parafil ropes; they are made as a bundle of parallel yarns, which are in turn made from about 1000 individual filaments. The filaments themselves have strengths of about $3500 \, N/mm^2$; the yarns have a strength of about $2900 \, N/mm^2$, and the ropes have a minimum strength of about $1930 \, N/mm^2$.

These effects are described by bundle theory, which accounts for these size effects. It is also possible to account for length effects in a similar way, by using weakest link theory. Both theories rely on the variability of strength of yarns and fibres and predict that the strength will reach an asymptotic value as the rope size gets bigger, and also as it gets longer. Figure 5.5 shows the variation in strength with rope size as measured in Type G Parafil ropes. Other tests have been carried out on much larger ropes (up to 1500 tonnes break load) and these lie very close to the

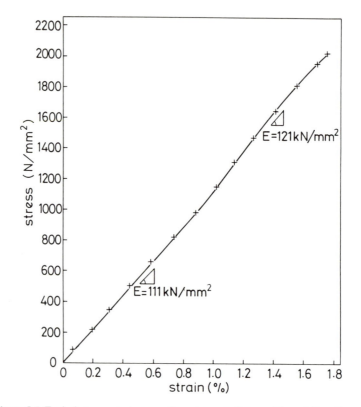

Figure 5.4 Typical stress–strain curve from tests on 60 tonne Type G Parafil ropes.

asymptotic value. Thus, for all practical rope sizes and lengths used in prestressing, the strength of the ropes can be taken as 1930 N/mm², measured over the cross-sectional area of the yarns.

5.5.2 Creep, relaxation and loss of prestress

Aramid fibres offer significantly lower creep than most other fibres used in rope making; indeed, for many rope applications, the creep is negligible. However, when used for prestressing concrete, engineers are interested in the creep as a proportion of the initial extension, since this governs the amount of prestressing force lost.

Total creep strains are of the order of 0.13%, which can be compared with a rope extension (when stressed to about 50% of its initial break load), of about 0.8%. Thus, we can expect to lose something like 16% of the initial prestress force in a Parafil tendon.

Figure 5.6 shows predicted stress relaxation figures for different ages

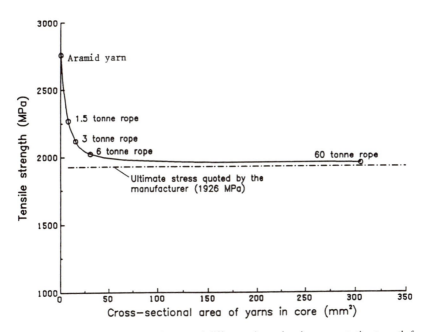

Figure 5.5 Measured strengths of ropes of different sizes, showing asymptotic strength for large ropes.

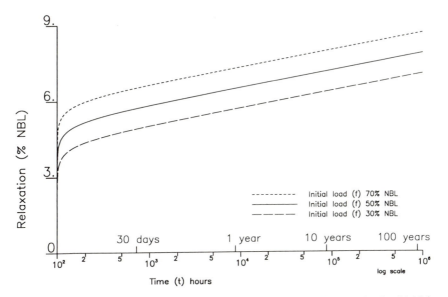

Figure 5.6 Stress relaxation predictions for Type G Parafil, from different levels of initial prestress, expressed as a percentage of the nominal break load (NBL) (Chambers data).

and for different initial stresses (expressed as percentages of the nominal break load, NBL), based on a series of rope tests. It can be seen that they agree quite closely with the limiting value given above.

The total loss of prestress force in a member prestressed with Parafil is very similar to that in a beam prestressed with steel. The losses due to the relaxation of the tendon are higher, as explained above, but this is compensated by reduced losses due to the shortening of the concrete. Kevlar yarns have a lower elastic modulus than steel (approx. 2/3), so that the loss of force in the tendon caused by a reduction in length of the concrete is about two-thirds of that in steel tendons. This will be true for losses caused by elastic shortening of the concrete and also for losses due to creep of the concrete. Friction losses are of a similar order to those with steel tendons but, especially when using external tendons, where friction losses occur at discrete points where the tendon is deflected, it is probably worth wrapping the tendon in PTFE or using a lubricant to reduce the friction further.

As with all loss calculations, the total losses depend on details of the design, which will differ for structures designed with steel or Parafil tendons, but for most cases the various effects cancel one another fairly closely.

5.5.3 Stress rupture

Stress rupture, or creep rupture as it is sometimes known, is the name associated with failure caused by a material creeping until it breaks. This is not normally a problem in steels, except at high stresses or high temperatures, but it is likely to be a governing criterion for the long-term use of most systems that rely on new materials.

Stress rupture is clearly related to creep and relaxation. At higher stresses, materials creep more and fail in a shorter period of time than at low stresses. There are strong theoretical arguments, related to the activation energy of the creep process, why there should be a linear relationship between the applied stress and the logarithm of the lifetime of the material. This is indeed observed in tests on both Parafil ropes and on Kevlar yarns. Figure 5.7 shows values of lifetimes as measured in tests on Parafil ropes and compares them with theoretical predictions based on tests performed on Kevlar 49 and epoxy bars. The results have been normalised with respect to the short-term strength, because of the bundle theory effects described above.

Statistical analyses have been carried out on these data and it is predicated that a rope loaded to 50% of its short-term strength will have a 1.4% chance of failing if the load is maintained continuously for a period of 100 years. This prediction is based on extrapolation of tests carried out at ambient temperature for periods of about 4 years, and from con-

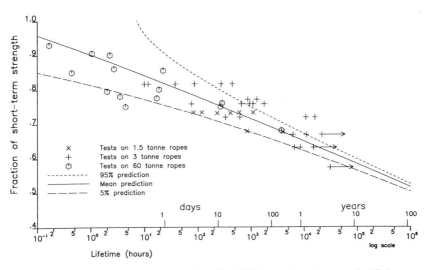

Figure 5.7 Stress rupture test results and predicted lifetimes, based on tests by Guimaraes and Chambers.

sideration of tests carried out for shorter periods at elevated temperatures (which can be related to those at ambient temperature via the activation energy). At the moment, tests are underway with ropes loaded by dead weights to produce failures in ropes after periods in the 5–10 year range. These will give engineers more confidence in the extrapolation up to structural lifetimes.

Work is currently underway identifying the cumulative damage rule that must be applied if a rope is subject to varying loads. The most likely rule appears to be one where the rope sustains stress rupture damage as a linear proportion of the lifetime that it spends at each load. This allows the stress rupture lifetime to be calculated where the load is reducing as the prestress force drops off because of concrete creep and tendon relaxation.

One point needs to be made about these results. The stress rupture lifetime relates to loads applied continuously; it does not mean that the short-term strength is reduced by the same extent. The strength retention observed in a rope that has been subjected to a load for half of its stress rupture lifetime would be virtually unchanged from the short-term strength. This has important implications for prestressing concrete. The initial prestressing force can be chosen on the basis of the long-term stress rupture of the tendon, taking due account of the relatively short period of time the rope spends at a higher force before creep and relaxation have occurred. But the force in the tendon then changes very little due to live load effects, other than very occasional excursions when the structure is

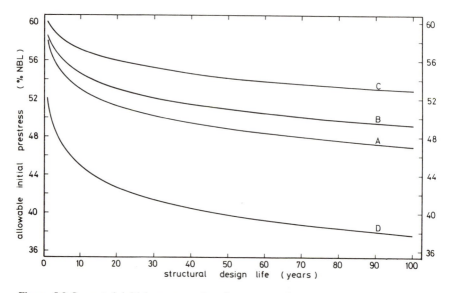

Figure 5.8 Suggested initial prestress, based on assumed losses of force due to stress relaxation and creep of concrete, to give a 10^{-6} probability of failure due to stress rupture (based on Chambers test results). For definition of curves A–D see text.

overloaded. On these occasions, the tendon will still have virtually its full strength.

These ideas can all be combined. Figure 5.8 shows allowable initial prestressing forces in a beam to give a 10^{-6} chance of failure due to stress rupture for a given design life. Curve D shows the maximum allowable force for a constant load, but curves A, B and C show the situation for minimum, typical and maximum prestress losses, respectively. For the case of typical losses, an initial prestressing force that is about 10% higher can be allowed, since the force will subsequently reduce, thus reducing the stress rupture damage that is taking place.

5.5.4 Durability

The tendons can be expected to have high durability in normal environments. Kevlar is degraded by ultraviolet light, but this is shielded by the sheath and is not a problem. Kevlar fibres also suffer hydrolytic attack by strong acids and alkalis, but the tendons would not be bonded to the concrete, so the fibres will not come into contact with the alkaline concrete. In any event, the sheath will act as a barrier to ingress of chemicals. DuPont have reported that Kevlar is not degraded by either fresh or salt water at normal pH levels.

There are potential concerns about fire and vandalism with these materials. Kevlar does not burn; it has very similar fire resistance properties to Nomex, which is chemically very similar and is widely used for protective clothing by firemen and racing drivers. It decomposes at a temperature of about 450°C, and loses about half of its strength at about 250°C. It has very low thermal conductivity, so in large ropes, the central core of fibres will not lose strength as quickly as the outer fibres. Nevertheless, it is likely that some attention should be paid to protecting Parafil from fire, especially when used as external tendons.

There is also the potential problem of mechanical damage, especially due to vandalism. Prestressing tendons are not normally accessible to the general public, but in cases where they can be reached, consideration should be given to putting the tendon inside a casing of some sort.

5.5.5 Fatigue

The fatigue characteristics of aramid fibres are very good. The resistance of Kevlar to tension-tension fatigue is better than that of steel, and is probably due to cumulative damage due to stress rupture, rather than simply the number of cycles. When 'fatigue' failures of Kevlar do occur, they are normally due to fretting of fibres over one another. This can only occur at the terminations, or at loading points, and the variation in force in prestressing tendons, especially when unbonded, is extremely low. Thus, it is not believed that fatigue is a problem in prestressing applications.

5.6 Structural applications other than prestressing

Because the ropes can be made in almost any size and with efficient terminals at the ends, it is possible to use them in a variety of ways. Several novel applications have been made, where the various properties of Parafil have contributed significantly to the success of the scheme.

5.6.1 Bicentennial tent

During the Australian Bicentennial celebrations in 1988, a touring exhibition was mounted, which consisted of a series of tented exhibition stands mounted on trucks (Figure 5.9). These were moved around the country and erected many times at different locations. Parafil ropes were used as the main supporting cables for the tents, and also as the tensioning elements around the edges of the tents. They were lighter than the equivalent steel cables and stood up well to the rigours of repeated assembly and disassembly.

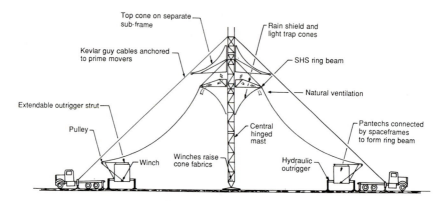

Figure 5.9 Tent for Australian bicentennial touring exhibition.

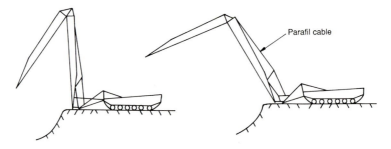

Figure 5.10 Scissoring action of tank-launched bridge, controlled by Type G Parafil (Defence Research Agency).

5.6.2 Tank bridge

The British Army uses an armoured deployable bridge system mounted on a tank chassis. This uses a rope to deploy the bridge in a scissoring action (Figure 5.10); the available lever arm is small, so the forces that have to be carried are high. The rope must be stiff as well as strong, as it controls the accuracy with which the bridge can be placed. Conventional steel rope is very heavy and is awkward to both carry and stow. To overcome these problems, the Defence Research Agency carried out tests using Parafil ropes. They are much easier to handle and are just as effective in operating the bridge; future versions of the bridge are likely to incorporate Parafil scissoring ropes.

5.6.3 Bus station roof, Cambridge

A small bus station was completed in Cambridge in 1991; this has a roof supported by Parafil ropes with a 7 m cantilever (Figure 5.11). There are

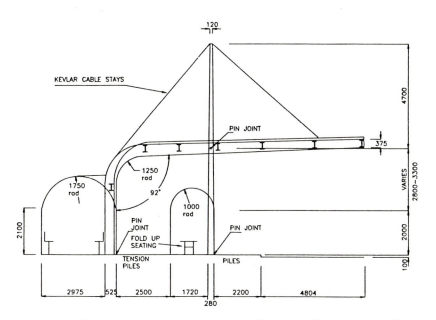

Figure 5.11 Cable-stayed roof on bus station at Cambridge (Cambridgeshire County Council).

four masts, each supporting a pair of forestays and a pair of backstays. Although designed primarily to resist snow loading, the stays are permanently stressed to ensure that the roof remains stiff even when wind loads cause uplift. The structure was designed by Cambridgeshire County Council and the ropes were fitted with terminals and preloaded in the Engineering Laboratories at the University of Cambridge.

5.6.4 Aberfeldy bridge, Scotland

The western world's first all-plastic bridge has just been completed at Aberfeldy in Scotland. It combines a deck and towers made from lightweight glass-reinforced plastic pultrusions (developed by Maunsell Structural Plastics Ltd and manufactured by GEC Reinforced Plastics) with stay cables made from Parafil. The bridge carries a footpath linking two halves of a golf course across the River Tay, with a clear span of 64 m (Figure 5.12). The only non-plastic components are concrete in the foundation and some connecting pieces between the deck and the cable terminations to distribute the concentrated load.

The bridge was built by students from Dundee University; the pultrusions were assembled on scaffolding on shore and then launched across the river, supported on a cats-cradle of cables made up from the per-

Figure 5.12 Aberfeldy bridge, Scotland. The deck and towers are glass-reinforced plastic; the cables are Type G Parafil ropes (reproduced by kind permission of Maunsell Structural Plastics).

manent Parafil stay cables and some temporary cables. The lightness of all the components meant that no cranes were needed during the erection.

5.7 Prestressing applications

5.7.1 Thorpe Marsh Power Station

Thorpe Marsh electricity generating station in the north of England was one of a series built in the 1960s to use coal mined locally. It has six large cooling towers, of which three were recently found to have large cracks at the top; this left them in a very unstable condition. Demolition would have kept the station out of service for a considerable period, but it was decided that the towers could be repaired by circumferential prestressing after injecting the cracks with resin; Parafil ropes were used for this application (Figure 5.13). The prime benefits were the resistance to corrosion and the light weight, which meant that the prestressing could be carried out by steeplejacks carrying coils of cable up the towers. They could work their way round the towers, installing the cables as they went, before stressing the cables one-by-one. The alternative, using steel cables, would have meant assembling a net at ground level, and lifting it up by means of cranes and then adjusting the lengths of all the elements; a much more complex operation. After several years in operation, the Parafil ropes are performing well.

Figure 5.13 Cooling towers at Thorpe Marsh electricity generating station, Doncaster, circumferentially prestressed with Parafil ropes.

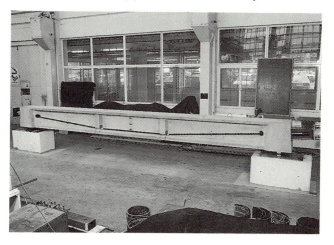

Figure 5.14 Beam (8 m long) prestressed externally with two 60 tonne Type G Parafil ropes.

5.7.2   Beam tests at Imperial College

Tests have been carried out on two beams prestressed with Parafil to demonstrate the feasibility of producing structural elements in this way. Two designs were produced; the first had a single, straight unbonded tendon, contained within a duct on the centreline of a simple I-beam, while the second had two external deflected tendons, one on each side of a T-shaped cross-section (Figure 5.14). In each case, the tendons were

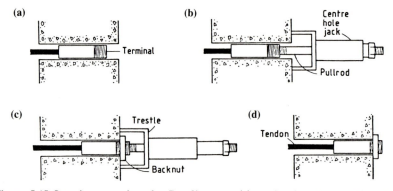

Figure 5.15 Stressing procedure for Parafil ropes: (a) tendon installation; (b) jack attachment; (c) prestressing force application; (d) final arrangement.

Type G Parafil with a nominal break load of 60 tonnes, prestressed to about 50% of their short-term strength.

For an internal tendon, the terminals have to be fitted before the rope is placed in position in the beam; since the terminals are too large to pass through the duct, this is built up around the tendons. In the second beam, the tendons were to be placed outside the concrete, so there was no need to assemble the rope in a duct prior to casting. Holes were formed in the thickened end blocks to receive the rope terminations, by casting-in plastic pipes.

The principles of the stressing procedure are shown in Figure 5.15. The tendon is placed in the structure and a pull-rod fitted to the internal thread of the termination. The pull-rod is then passed through the centre hole of a hydraulic jack and secured by means of a nut. The jack is held away from the beam by means of a trestle, which allows access to the terminal to secure the back-nut. Force is applied by the jack, which brings the terminal just outside the face of the concrete; the back-nut can then be fitted to lock the tendon in position in its stressed state. The jack, trestle and pull-rod are removed, and a security cap fitted to prevent dirt and debris getting into the termination. This would also serve to contain the anchorage in the unlikely event of a rope failure.

Measurements of the forces in the second beam showed that the coefficient of friction was about 0.32, which is slightly higher than would be expected with steel tendons, but could be brought down by a better selection of sheath and deflector material. Measurement of the force in the tendons, after the force had been transferred from the jack to the permanent back-nut, indicated that no loss of prestress occurred at this stage.

5.7.3 Load–deflection behaviour

Both beams were tested in four point bending rigs, with loads applied by hydraulic jacks. The beams were taken through several elastic loading cycles; the second beam was kept under sustained load for 42 days to monitor the effects of creep and relaxation.

The relationship between the applied load and the deflection at the centre of the second beam is shown in Figure 5.16. On the application of the load, the response is almost linear, with the portions of the curves corresponding to loading and unloading being parallel. The instantaneous camber produced by the prestressing is indicated by the horizontal part of the curve at zero load. The increase of deflection due to the effects of shrinkage and creep of concrete after 42 days was 59% of the instantaneous deflection caused by the applied load. This figure is not affected by relaxation of the tendon, the increased deflection being due to loss of stiffness of the concrete.

The total loss of prestress in both tendons of the second beam, due to shrinkage and creep of concrete and due to stress relaxation in the tendons, is shown in Figure 5.17. The tendons were tensioned initially to approximately 22% of their tensile strength, when the age of concrete was 10 days. Losses of 13% and 14% of the initial force were observed in the two tendons after 23 days, when the full prestressing force was applied. Over this period of time, the beam was subjected only to its own

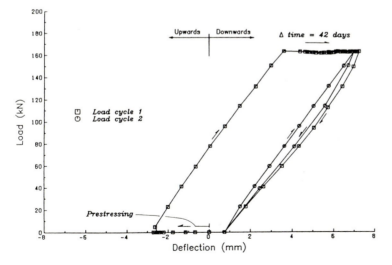

Figure 5.16 Applied load versus mid-span deflection for the second (8 m) beam, showing camber due to prestress, elastic response due to load and extra deflection due to creep of concrete.

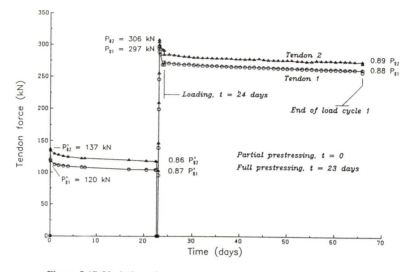

Figure 5.17 Variation of prestressing force with time in an 8-m beam.

weight. Forty-three days after the application of the full prestressing, the losses of prestress under service load were 12% in tendon 1 and 11% in tendon 2. It can be seen in the figure that most of the losses occurred within the first day after prestressing. From then on the curves show a very low rate of loss. These figures are very similar to losses to be expected in steel tendons, in accordance with the comments made earlier.

Ultimate load tests were carried out on both beams, which responded as expected. After passing the cracking load, the stiffness reduced considerably; when unloaded from the cracked (but still elastic) state, the stiffness remained lower until the cracks had closed up but the full elastic stiffness was recovered and there was virtually no permanent set.

When loaded until failure, both beams showed considerable curvature at virtually constant load, with large cracks forming in the bottom of the beam. Failure occurred in both beams by crushing of the top flange. Figure 5.18 shows the load–deflection curves for the second beam; the results for the first are similar. Even though failure of the top flange concrete precipitated the final failure, there was a lot of warning of failure as the cracks opened up.

There were slight differences in the final failure mode of the two beams which cast important light on the behaviour of unbonded and external tendons. In both cases, the top flange failed by crushing, but in the first beam, as the tendon was constrained in the bottom flange, the beam did not completely collapse. The compression zone passed down through the web and into the top of the bottom flange, with a consequent reduction in

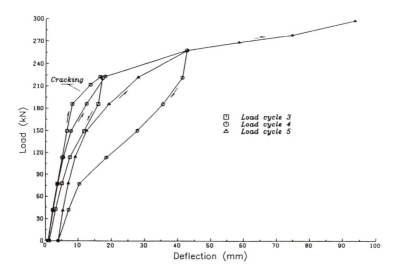

Figure 5.18 Load–deflection curve for an 8-m beam when loaded to failure. Note the plateau corresponding to crack opening.

load. However, the bottom flange did not fail, remaining axially pre-stressed. After the test, the tendon was found still to be carrying a significant force.

In the second beam, the tendon was outside the bottom flange, which could thus deflect while leaving the tendon in its original position relative to the ends of the beam. The beam thus failed suddenly and completely, with a total loss of prestress.

The results obtained in these tests were very similar to those that would have been expected with unbonded steel tendons. There would be slight differences due to the different Young's modulus of the Kevlar, but the overall behaviour of the beam and tendon together, as a composite system, is not affected by the different material. Ductility and rotation capacity of such beams should not be a problem, as the concrete can crack and slide relative to the tendon, without causing significant prob-lems. The new design rules for unbonded tendons in bridges are awaited from the Department of Transport with interest; the principles (at least) should be applicable to Parafil as well as to steel tendons.

5.7.4 Reasons for not bonding to concrete

Aramids, like carbon fibres and glasses, exhibit brittle behaviour when tensile loads are applied to them, which means that they are very sensitive to applied strains which exceed the design values. This is in contrast to steel, where the plateau on the stress–strain curve means that the tendon

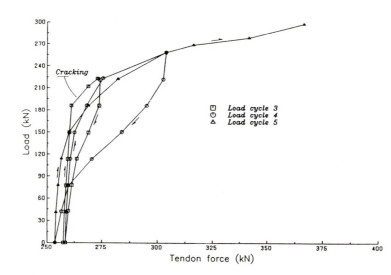

Figure 5.19 Change in tendon force in an 8-m beam when loaded to failure. Note the relatively small change in force as the tendon was unbonded.

can absorb high strains locally with no significant problems other than a permanent set.

We must therefore be very careful when deciding whether to bond these materials to concrete. In the vicinity of cracks, the local strains are very high; indeed, if we have perfect bond between steel and concrete, they are infinite. Thus we might expect that beams prestressed with any of these materials, if they are bonded to the concrete, will fail by local snapping of the tendon, with no possibility of redistribution of load, or of plastic deformation to accommodate the strain. There may be some redistribution of load due to bond failure, but this will not significantly alter the problem.

Thus, beams prestressed with new materials should be designed with unbonded tendons. This has some implications for design procedures, as there will be relatively little increase in tendon force as the beam is loaded.

This effect was observed in the beam tests; Figure 5.19 shows the tendon force during the ultimate load cycles on the second beam, from which it can be seen that fairly small changes in tendon force occurred, even though the concrete was significantly cracked, because the tendon could slide relative to the concrete. There is thus very little chance of snapping the tendon. The change in resistance to external bending moment is almost exclusively due to an increase of the lever arm between the internal compression and tension forces whose magnitudes remain relatively unchanged.

Table 5.2 Indicative sizes of Parafil ropes (data from VSL International)

Nominal break load (kN)	Rope diameter (mm)	Rope weight (kg/m)	Termination	
			Length (mm)	Diameter (mm)
1000	40	0.42	520	130
2000	55	0.72	650	180
3000	67	1.03	950	220

Parafil ropes, with their polyethylene sheath, cannot be bonded to concrete. Even if they are cast in place, there will be slip between the tendon and the sheath and creep of the sheath itself.

5.8   Link with VSL International

Parafil ropes have now moved from the development and prototype stage to fully fledged prestressing tendons with an agreement between Linear Composites Ltd and the VSL International group, who are one of the world's leading suppliers of prestressing systems. Engineers will now be able to design structures with Parafil prestressing, in the knowledge that jacking systems and specialist assistance will be available at the time of installation. The jacking system used will follow the principles described above with special fittings to allow the use of existing VSL jacks. Typical sizes of prestressing tendons are shown in Table 5.2. VSL have recently introduced polyethylene duct systems for conventional prestressing tendons; combining these with Parafil will completely remove the danger of corrosion of the prestressing system.

5.9   Predictions for future prestressing systems

It is expected that beams, both in bridges and buildings, will be prestressed with Parafil. The tendons will either lie outside the concrete, or unbonded in ducts within the concrete. No account will be taken of increased forces in the tendon due to live load.

The results of the Imperial College tests show that basic design principles for prestressed concrete do not need altering radically; the following are points which a designer should take into account when designing a beam with Parafil tendons:

1. The tendon should be pretensioned, with the terminals in place, to a load level in excess of that expected during both the initial

stressing operation and the service life of the structure. This will have the effect of ensuring that the terminal spike is properly bedded and will also give a check on the tendon length before being placed in the structure. It is normal practice, according to the manufacturer's instructions, to pretension ropes to 60% of the nominal breaking load prior to use, whenever possible. These ropes, when used in conventional rigging arrangements, are normally stressed to much lower load levels than those in use in prestressing tendons; in these cases, 60% is perfectly adequate as a pretensioning load. However, in prestressing tendons, where high force levels are normal, a higher pretensioning level may be needed to ensure adequate bedding of the termination.

2. Any deflector points should be properly flared to ensure no damage to the sheath during stressing operations; this should not be difficult to arrange if taken into account at the design stage.

3. The coefficient of friction between the tendon and the duct (or the deflector) should be reduced wherever possible. This may mean undertaking some studies of friction coefficients between various possible sheathing materials and alternative duct materials. Alternatively, coating materials, such as PTFE or nylon tapes, might be considered.

4. The working load design of prestressed concrete beams should be based on allowable stress limits taking account of the design prestressing force, after allowing for losses, and the ultimate strength of the section should be based on the assumption that only minimal increases of force take place due to geometry changes as the beam deflects.

5. The compression zone of the concrete should be provided with confining reinforcement to increase the ductility of the concrete in that area.

6. If the tendons are external to the concrete, they should pass through loose rings so that, in the event of failure, the tendons are forced to deflect with the beam. This will ensure that failure occurs in the more controlled manner of the first beam.

5.10 Conclusion

It is clear that Parafil will start to find more widespread use as a non-corroding prestressing tendon. Repair of structures by the use of external tendons is already taking place and will become more common; the use of unbonded, replaceable tendons is likely to become the norm for all structures in the near future. Structures, such as water towers, which often have poorly protected steel prestressing and a high incidence of corrosion, are currently being studied with a view to their repair with external Parafil tendons. A number of bridges with suspect prestressing tendons are also being identified by the current bridge assessment programmes; these would make useful demonstration sites for new materials as the new tendons would be adding an extra margin of safety, rather than providing the primary stressing.

Offshore, as exploration for oil and other minerals moves into ever deeper water, the arguments for using mooring lines with almost neutral buoyancy become more persuasive. Structures can be moored in 300 m of water using steel, but not in 3000 m. Virtually all the major oil companies have conducted studies into the use of lightweight mooring lines; when economics dictate that such structures be built, Parafil ropes, or similar systems, will undoubtedly be used.

Similarly, as bridge spans increase, the use of lightweight stiff materials becomes more economic. The excellent fatigue behaviour will also be seen to be important. The Eurobridge proposal to cross the English Channel with seven spans of 4.5 km was probably 20 years ahead of its time, and had some conceptual flaws. Nevertheless, such large spans are only going to be possible if new materials are used.

Other applications will make use of the non-magnetic nature of the material; applications such as de-Gaussing facilities for ships, or as strength elements in members carrying important communications (such as railway signalling and control equipment), can also be envisaged.

Bibliography

The following bibliography is not exhaustive, but includes major works relating to the properties of Parafil, and all works from which data given here have been drawn.

1. Kingston, D., Development of parallel fibre tensile members, *Symp. on Engineering Applications of Parafil Ropes*, Imperial College, London, 1988.
2. Chambers, J.J., Parallel-lay aramid ropes for use as tendons in prestressed concrete, Ph.D. Thesis, University of London, 1986.
3. Guimaraes, G.B., Parallel-lay aramid ropes for use in structural engineering, Ph.D. Thesis, University of London, 1988.
4. Burgoyne, C.J. and Chambers, J.J., Prestressing with Parafil tendons, *Concrete* 19(10) (1985) 12–16.
5. Burgoyne, C.J., Structural uses of polyaramid ropes, *Construct. Building Mater.* 1 (1987) 3–13.
6. Chambers, J.J. and Burgoyne, C.J., An experimental investigation of the stress–rupture behaviour of a parallel-lay aramid rope, *J. Mater. Sci.* 25 (1990) 3723–3730.
7. Burgoyne, C.J., Properties of polyaramid ropes and implications for their use as external prestressing tendons, in *External Prestressing in Bridges*, eds. A.E. Naaman and J.E. Breen, American Concrete Institute, SP-120, Detroit, 1990, pp. 107–124.
8. Hobbs, R.E. and Burgoyne, C.J., Bending fatigue in high-strength fibre ropes, *Int. J. Fatigue* 13 (1991) 174–180.
9. Burgoyne, C.J., Guimaraes, G.B. and Chambers, J.J., Tests on beams prestressed with unbonded polyaramid tendons, Cambridge Univ. Eng. Dept. Tech. Report CUED/D, Struct/TR. 132, 1991.

10. Guimaraes, G.B. and Burgoyne, C.J., The creep behaviour of a parallel-lay aramid rope, *J. Mater. Sci.* **27** (1992) 2473–2489.
11. Guimaraes, G.B. and Burgoyne, C.J., Repair of concrete bridges using Parafil ropes, US European Workshop on Rehabilitation of Bridges.
12. Burgoyne, C.J., ed., *Proc. Symp. on Engineering Applications of Parafil Ropes*, Imperial College, London, 1988.
13. Burgoyne, C.J., Structural applications of Type G Parafil, *Symp. on Engineering Applications of Parafil Ropes*, Imperial College, London, 1988.
14. Burgoyne, C.J., Polyaramid ropes for tension structures, *1st Int. Oleg Kerensky Memorial Conf. on Tension Structures*, London, 1988.
15. Burgoyne, C.J., Laboratory testing of Parafil Ropes, Les matériaux nouveaux pour la precontrainte et le renforcement d'ouvrages d'art, LCPC Paris, 1988.
16. Burgoyne, C.J. and Flory, J.F., Length effects due to yarn variability in parallel-lay ropes, MTS-90, Washington, DC, 1990.
17. Snell, M.B. and Taylor, R.M., The use of Parafil ropes in tank launched bridges, RARDE Div. Note EE/2/89.
18. Dean, B.K. and Wynhoven, J.H., Australian bicentennial travelling exhibition, *1st Int. Oleg Kerensky Memorial Conf. on Tension Structures*, London, 1988.
19. Richmond, B. and Head, P.R., Alternative materials in long-span bridge structures, *1st Int. Oleg Kerensky Memorial Conf. on Tension Structures*, London, 1988.
20. Burgoyne, C.J., Tests on beams prestressed with polyaramid ropes, *Proc. 1st Int. Conf. on Advanced Composite Materials in Bridges and Structures*, Sherbrooke, Quebec, 1992.
21. Amaniampong, G., Variability and visco-elasticity of parallel-lay ropes, Ph.D. Thesis, University of Cambridge, 1992.
22. Burgoyne, C.J., Should FRP tendons be bonded to concrete? *Int. Symp. on Non-metallic Reinforcement and Prestressing*, American Concrete Institute, 1993.

6 Glass-fibre prestressing system

R. WOLFF and H.-J. MIESSELER

6.1 Introduction

High performance glass-fibre composite materials, until recently almost exclusively used in the aerospace or car industries, are now also applied in the construction industry. The choice of different kinds of fibres, such as carbon, glass or aramid fibres, and of different kinds of resin, e.g. polyester, epoxy or vinyl, offers the possibility of producing fibre composite materials for a large range of applications.

Glass-fibre composite materials, especially used in the construction industry as prestressing elements, distinguish themselves by their tensile strength which is equal to or higher than that of high tensile steel. They have a high elasticity, an excellent corrosion resistance and there is the possibility of integrating optical fibre sensors for a permanent monitoring of fibre composite materials and the concrete structures themselves.

6.2 Material characteristics

6.2.1 Available dimensions and strengths

The glass-fibre composite bars employed to date comprise 68 vol.% glass fibres and 32 vol.% unsaturated polyester resin or epoxy resin. A total of 64 000 strictly uni-oriented E-glass fibres with a diameter of approx. 25 µm form a 7.5 mm diameter bar. The longitudinal tensile strength of this new composite material is similar to that of high-tensile prestressing steels [1]. A comparison of the stress–strain curve of glass fibre materials with that of comparable conventional ST. 1470/1670 prestressing steel is shown in Figure 6.1. Table 6.1 gives the main characteristic values of the material in comparison to the steels and composite bars comprising other fibres. These different materials are also classified in areas of application. The main distinctions between the newly developed glass-fibre prestressing tendons (HLV tendons) in comparison to prestressing steel are:

- the modulus of elasticity for HLV tendons at 51 000 N/mm^2 amounts to only one-fourth of that of steel prestressing tendons;
- the prestressing tendons show linear stress–strain behaviour until failure;

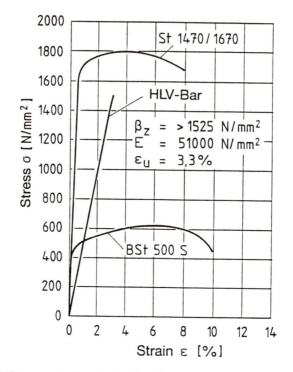

Figure 6.1 Stress–strain diagram of a glass-fibre bar in comparison with steel bars.

- low specific weight of $2.0\,\text{g/cm}^3$ compared with $7.85\,\text{g/cm}^3$ for pre-stressing steel;
- possibility of integrating sensors to create intelligent prestressing systems.

Whilst a linear and a plastic range can be differentiated with steel, the characteristic curve for glass-fibre composite material up to failure point complies precisely with Hooke's Law.

Other important properties of the glass fibre composite material are:

- the long-term strength amounts to 70% of the short-term strength;
- good resistance in aggressive environments;
- shows the same behaviour at high temperatures as prestressing steel;
- electro-magnetic neutrality.

The glass-fibre bar with polyester resin is provided with a purpose-developed polyamide coating. A special epoxy resin powder coating is used for glass-fibre bars with epoxy resin. The coatings act as a protection

Table 6.1 Material characteristics and comparisons

	Reinforcing steel BS500	Prestressing steel BS1470/1670	Polystal^R (68% glass fibres)	Arapree^R (Aramid fibres)	Carbon fibre composite material
Tensile strength (N/mm²)	>550	>1670	1670	1610	1700
Yield strength (N/mm²)	>500	>1470	–	–	–
Ultimate strain (%)	10	6	3.3	2.5	1.1
Modulus of elasticity (N/mm²)	210 000	205 000	51 000	64 000	146 000
Specific weight (g/cm³)	7.85	7.85	2.0	1.3	1.5
Fields of application	Reinforced concrete structures	Prestressed structures			Stay cables bracings

Figure 6.2 Time-dependent tensile strength of glass-fibre composite bars.

against chemical environments, for example against chloride and alkaline, or mechanical damage.

6.2.2 Long-term behaviour

Long-term behaviour under permanent static load is illustrated in Figure 6.2. Due to the relatively (but for this type of test not unusually large) dispersion of individual values, only a calculated value for the expected

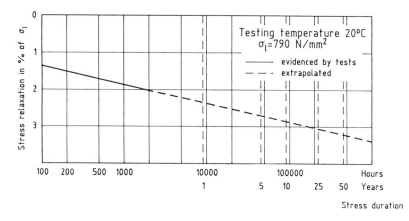

Figure 6.3 Relaxation of the glass-fibre composite bars.

time-dependent creep strength can be estimated. At present, a time-dependent tensile strength of 70% of the short-term strength is assumed. This value is divided by a safety factor of 1.42 in the design process.

If a bar is loaded to some value less than 70% of its short-term strength, then creep is of no significance. The load capacity of a bar loaded for a long period of time to, say, 40% will equal the short-term value.

The relaxation processes observed with composite glass-fibre bars are traceable to irregularities caused in manufacture, e.g. deviations from the strict axial orientation of the fibres within a bar. As with prestressing steel, stress losses resulting from relaxation show linear behaviour if plotted along a logarithmic time axis (Figure 6.3). It is thus possible to extrapolate data gained in short-term tests over longer periods. A charac-teristic of $\sigma_{z,t} = 3.2\%$ is derivable for an application as prestressing reinforcement for a time period of 5×10^5 h. A significant change in this behaviour is not to be expected even at higher temperatures (e.g. 50°C).

6.2.3 Dynamic behaviour

The influence of non-static loads was assessed in fatigue tests whereby, with respect to utilisation as prestressed reinforcement, only the range of increasing tensile stresses which includes stress generated by the pre-stressing force, is of significance. The results of such tests, which were carried out at a constant maximum stress limit to the extent of the working load, are shown in Figure 6.4. While in the case of prestressing steel it is permitted to define the resultant 2×10^6 variable stress com-ponent as the fatigue strength, in the case of composite fibre materials, failures still have to be expected even with higher load cycles. A possible consequence of this is that the load cycles to be expected for a structure

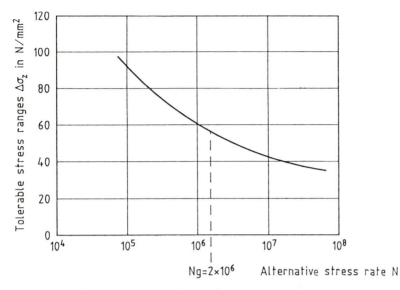

Figure 6.4 Fatigue behaviour (10% failure probability).

must be forecast and investigated in each individual case to determine whether the tolerable stress ranges are sufficient.

6.2.4   Influence of temperature

In fire tests on structural elements carried out at Brunswick University, an exposure time of over 100 min at 1000°C in a fire chamber could be achieved using the minimum concrete covering for the prestressing tendon, as laid down by ZTVK (Supplementary Technical Regulations for Bridges and Tunnels). Proof of fire classification F 90 demonstrates the glass-fibre material's applicability at high temperatures.

Glass fibres themselves are relatively insensitive to high or low temperatures. However there is partial loss of strength in the matrix sheath at approx. 250°C. The glass-fibre bars therefore have to be protected against overheating wherever the composite properties of the matrix are statically required, hence, particularly in the anchorage zone, by structural means, e.g. thicker concrete cover. The dependency on temperature of the tensile strength of glass-fibre composite material sections is shown in Figure 6.5, which shows that the response is similar to that of prestressing steels.

6.2.5   Behaviour under the effects of aggressive media

Reinforcement tendons for applications in prestressed concrete structures must be permanently resistant to the strongly alkaline reacting medium

$$\frac{\text{Ultimate tensile strength } B_z(T) \text{ at elevated temperatures}}{\text{Ultimate tensile strength } B_z(T_0) \text{ at ambient temperature}}$$

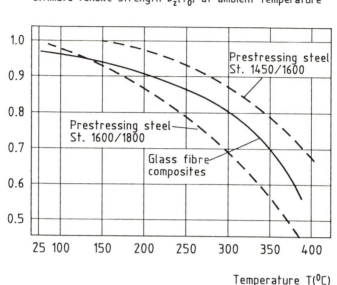

Figure 6.5 Tensile strength retention at elevated temperatures.

of concrete. Moreover, during prestressing they must withstand high mechanical loads without damage. Both requirements place high demands on the bar material. Although glass fibres are sensitive to many media due to their exceptionally large surface area relative to their volume, glass-fibre composites are employed successfully and comprehensively for the manufacture of pipes, tanks and installations in chemical plant construction, because these composite materials are more resistant to many corrosive influences than other materials. In a composite material, the glass fibres are protected by the resin matrix. However, a prerequisite for permanent protection is that strain remains low (0.1–0.2%) under operating conditions.

Prestressing reinforcement comprising composite glass-fibre bars, however, reach strains of >1.5% in their working state. These high strain values fundamentally change the behaviour of media influences. Under these conditions, the matrix material is no longer able to assure permanent protection of the glass fibres. During development of the glass fibre bars, it was therefore not possible to refer back to the experience gained in the construction of tanks. New concepts had to be put forward, both with respect to the basic materials and to the methods of testing. In this respect, particular reference should be made to test procedures for

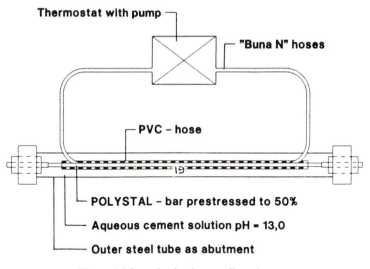

Figure 6.6 Investigation into media resistance.

the investigation of media influences subjected to simultaneous mechanical loads in compliance with practical conditions. These procedures substantially supported the specific development of utilisable composite glass-fibre bars for prestressing applications. While time-accelerated tests, which were performed under exaggerated conditions (temperatures 50–70°C, use of aqueous cement solution pH = 13) (Figure 6.6), provided relatively rapid conclusions with regard to the fundamental utility of a bar material, investigation of long-term media behaviour was effected in practice with the aid of so-called standard tests (Figure 6.7). The result of a sophisticated and arduous series of tests is a bar material with a coating consisting of a high-fill Polyamide 6, which on the one hand ensures media resistance and on the other withstands extreme frictional loads during prestressing.

6.3   The prestressing tendon and its anchorage

Being composite anisotropic materials, composite glass-fibre materials only tolerate transverse pressure up to 10% of their longitudinal tensile strength. For this reason, completely new solutions had to be found in the field of anchorage engineering. Thus, the absence of cold workability prevents the utilisation of upset heads, rolled-on threads or even the utilisation of steel wedges which 'bite' directly into the 'soft' composite glass-fibre bar material. The relatively low interlaminar shear strength of

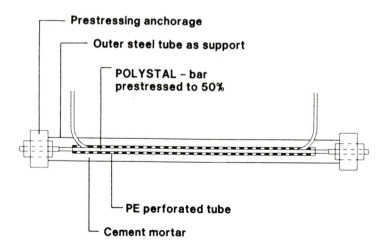

Figure 6.7 Investigation of resistance to media by means of standard tests.

the resin matrix requires a comparatively larger anchor length than would be the case with steel prestressing tendons.

On the basis of experience gained at Stuttgart University [2, 3] where clamping plate anchors had been used with batch-produced small rectangular bars, the round 7.5-mm diameter glass-fibre bars were tested by using modified clamping plate anchors [4]. The bars had been produced by a continuous manufacturing process since 1978. These experiments, in which the round bars were secured between semi-circular grooves in the clamping plates, did indeed show satisfactory results but also clearly demonstrated the limits of economical application using round bars. The development of a tubular grouted anchor at Strabag Bau-AG heralded a breakthrough for the anchorage of high performance glass-fibre composite tendons.

The composite tendon is grouted in a profiled steel tube with a synthetic resin specially developed for this purpose. The use of these grouted prestressing tendons covers the entire spectrum of light- and medium-weight prestressing tendons and the entire sphere of soil and rock anchoring (Figure 6.8).

The types of anchors developed up to the point of application maturity will be employed for the anchorage of prestressing tendons in prestressed concrete structures, and also for airside anchorage with soil and rock anchors. The anchorage technology of conventional steel soil anchors on the soilside can be adopted due to the relatively good bonding property of the bar material with the surrounding concrete medium (comparable with profiled prestressing steels).

Figure 6.8 Grouted anchorage of a high performance composite tendon.

6.4 Applications

6.4.1 'Lünen'sche Gasse' bridge

As early as 1980, a small bridge, 'Lünen'sche Gasse', was erected by the joint venture HLV Elements (Strabag Bau-AG, Bayer AG) as a trial structure within the scope of the Road, Bridge and Tunnel Department of the City of Düsseldorf. It was provided with limited prestressing without bond by means of 12 composite glass-fibre prestressing tendons and monitored by way of continual stress force measurements. Having finally been dismantled after 5 years for materials investigations, it was replaced by a new generation of prestressed tendon. The results confirmed the preceding laboratory tests as expected. The grouted anchorage for the round bars was then further developed up to application maturity based on the four anchorages employed (Figure 6.9).

6.4.2 Ulenbergstrasse bridge

Following the successful experience with the Lünen'sche bridge pilot project, the Road, Bridge and Tunnel Department of the City of Düsseldorf, which had already foreseen the new development, once again decided that this bridge, which is part of a heavily trafficked road, should be built

Figure 6.9 Lünen'sche Gasse bridge—overall view.

with the innovative prestressing system. Opened to traffic in 1986, the Ulenbergstrasse bridge represents a milestone in the continuous progress in the development of glass-fibre prestressing technology. Countless initial trials with the new bar material were carried out both in the laboratories of the joint venture partners and at notable German universities, so that finally all the knowledge gained could be concentrated in the bridge structure.

The Ulenbergstrasse road bridge (load classification 60/30 as per DIN 1072) is a two-span, solid slab bridge with span widths of 21.30 m and 25.60 m. The 1.44 m high and 15.0 m wide superstructure received limited longitudinal prestressing with a total of 59 HLV prestressing tendons (each with a working load of 600 kN) and was subsequently grouted with a synthetic resin mortar specially developed for this purpose [5, 6]. Measurements taken during the work, particularly tensioning paths during prestressing and deflection after handover to traffic, conformed well with the previously calculated values (Figures 6.10, 6.11).

6.4.3   Pedestrian bridge at Berlin-Marienfelde

The range of the prestressed bridge structures was extended by a further innovation in 1988. The Berlin-Marienfelde pedestrian bridge was not only distinguished by the novel prestressing method but was also the first

Figure 6.10 Ulenbergstrasse road bridge—overall view.

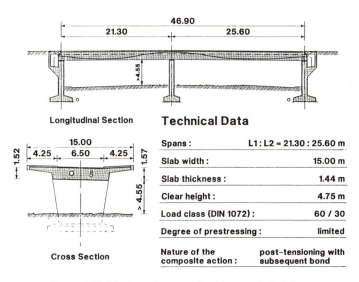

Longitudinal Section

Cross Section

Technical Data

Spans :	L1 : L2 = 21.30 : 25.60 m
Slab width :	15.00 m
Slab thickness :	1.44 m
Clear height :	4.75 m
Load class (DIN 1072) :	60 / 30
Degree of prestressing :	limited
Nature of the composite action :	post-tensioning with subsequent bond

Figure 6.11 Ulenbergstrasse road bridge—technical data.

building structure in Germany since the war to be executed with external prestressing. The superstructure consists of a two-span, 1.10 m high, TT-beam with span widths of 22.98 m and 27.61 m and is furnished with partial prestressing with 7 HLV prestressing tendons. The tendons run externally between the two main bridge beams, are diverted at two points on each span and upwards along the central pier with the aid of guide saddles, and anchored through the two cross-beams at the beginning and end of the bridge.

After completion, the bridge was loaded by the placing of 5 layers of 250 concrete slabs each weighing 1 tonne (corresponding to twice the total traffic load). This trial came within the scope of a research project and was accompanied by a comprehensive measuring programme in which changes in the prestressing force in the tendons and the deflection in the midspan were constantly controlled [7] (Figures 6.12, 6.13).

6.4.4   Schiessbergstrasse bridge

In the course of connecting services to the Bayer multi-storey car park, the Schiessbergstrasse to the north of the Bayer plant in Leverkusen was elevated to the second level. The three-span road bridge (load class 60/30) with span widths of $2 \times 16.30 \, m^2$ and $1 \times 20.40 \, m^2$ and a slab thickness of 1.10 m has limited prestressing with 27 post-bonded glass-fibre prestressing tendons with a working load of 600 kN each. The handover to traffic was in November 1992 (Figures 6.14, 6.15).

6.4.5   Nötsch bridge in Kärnten, Austria

The Nötsch bridge is the first bridge in Austria with glass-fibre prestressing tendons. The triple-span road bridge (bridge class I, Austrian Standard B 4002) with two span widths of 13.00 m and one of 18.00 m and a slab thickness of 0.75 m, is furnished with limited prestressing comprising 41 glass-fibre prestressing tendons, with post-bonds also with a working load of 600 kN each. The handover to traffic was in May 1992 (Figures 6.16, 6.17).

6.4.6   Brine tank at Bayer, Dormagen

Brine tank covers at Bayer's Dormagen plant were heavily affected or already destroyed by vapours with a high chloride content as a result of the highly corrosive environment. The two-span beam structures, which must withstand a load in compliance with bridge classification 60 according to DIN 1072, were replaced by a prestressed, prefabricated structure. Ulenbergstrasse type HLV prestressing tendons with a working load of 600 kN were also used in this instance. The excellent resistance of the glass fibres to an aggressive environment was decisive in the choice of this material. Previous observations had shown that use of this corrosion resistant prestressing reinforcement could considerably increase the life of building components subjected to such heavy demands.

Figure 6.12 Marienfelde bridge—overall view.

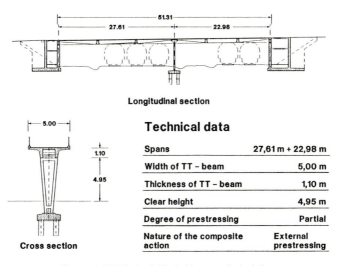

Longitudinal section

| 5.00 | | |

Cross section

Technical data

Spans	27,61 m + 22,98 m
Width of TT – beam	5,00 m
Thickness of TT – beam	1,10 m
Clear height	4,95 m
Degree of prestressing	Partial
Nature of the composite action	External prestressing

Figure 6.13 Marienfelde bridge—technical data.

6.4.7 *Mairie d'Ivry metro station (Paris)*

Rehabilitation of the Mairie d'Ivry metro station in Paris impressively established the wide range of application possibilities for HLV prestressing tendons. As a result of excavation directly adjacent to one side of the metro station, considerable cracking over a length of about 110 m had occurred in the approximately 70-year-old concrete vault arch owing to lateral load release. The client envisaged the installation of a prestressed

Figure 6.14 Schiessbergstrasse bridge—overall view.

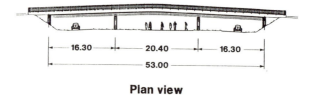

16.30	20.40	16.30

53.00

Plan view

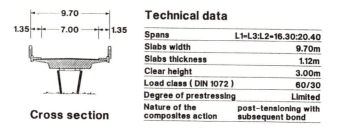

Cross section

Technical data

Spans	L1=L3:L2=16.30:20.40
Slabs width	9.70m
Slabs thickness	1.12m
Clear height	3.00m
Load class (DIN 1072)	60/30
Degree of prestressing	Limited
Nature of the composites action	post-tensioning with subsequent bond

Figure 6.15 Schiessbergstrasse bridge—technical data.

tie-rod as a rehabilitation and safety measure. The required prestressing of the arch walls was achieved by the installation of 36 HLV prestressing tendons and the rehabilitation measures were then executed as required. The electro-magnetic neutrality of the bar material proved to be most advantageous here. The tendons were prefabricated in the R&D Department laboratory and the prestressing works were executed by the same team.

Figure 6.16 Nötsch bridge—overall view.

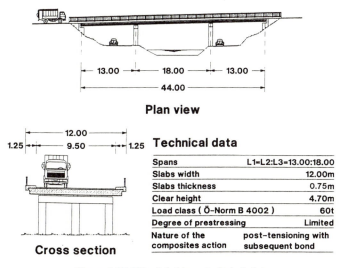

← 13.00 → ← 18.00 → ← 13.00 →

← 44.00 →

Plan view

← 12.00 →

1.25 ← 9.50 → 1.25

Cross section

Technical data

Spans	L1=L2:L3=13.00:18.00
Slabs width	12.00m
Slabs thickness	0.75m
Clear height	4.70m
Load class (Ö-Norm B 4002)	60t
Degree of prestressing	Limited
Nature of the composites action	post-tensioning with subsequent bond

Figure 6.17 Nötsch bridge—technical data.

6.4.8   Heydau monastery

During the rehabilitation of the Heydau monastery in Altmorschen, piers were anchored with masonry bolts to the corresponding masonry walls. Prestressing of the masonry anchors was required. Moreover, anchorage of the masonry bolts in the masonry itself was required so that visual damage to the historical front faces could be avoided. The chemical composition and the moisture condition of the masonry complicated the use of conventional anchorings with steel bolts. Therefore a safety concept on the basis of glass-fibre masonry anchors grouted with resin mortar was developed.

The tensile force of the prestressed masonry anchor is transferred to the sandstone blocks over a length of approximately 120 mm. With an applied prestressing force of 15 kN per anchor, there is a safety margin of 2.5 between the working conditions and failure.

6.5 Sensor technology

In order to monitor buildings in a useful way, it is necessary to develop appropriate sensors that are able to guarantee reliable measured values over a long period of time. Nowadays, intelligent processing of the high quantity of measured values is not a problem due to the availability and efficiency of personal computers.

The application of strain gauges on prestressing steel or the measurement of prestressing forces with the aid of load cells is not possible in the case of prestressing with post-bonded tendons. Moreover, it is not a durable solution in the case of prestressing without bond. Only the application of fibre composite materials facilitates permanent control of the prestressing element over its entire length by means of the integration of copper wire sensors or optical fibre sensors. Even the monitoring of each individual bar is possible. The sensors are already integrated into the tendon during its fabrication. The sensors indicate the integrity of the tendon or they locate any damage [8–11].

Besides the monitoring of prestressing elements, the observation of the stress–strain behaviour of concrete in the zone subject to tensile forces is very important. Therefore, the tensile zones above piers and spans are monitored permanently with integrated optical fibre sensors with a measurement accuracy of ±0.2 mm. With specially developed optical strain sensors, it is possible to measure the influence of loadings with a short-term accuracy in the range of micrometres.

6.5.1 The crack detection sensor

In this type of sensor, one or several optical fibre sensors are stranded with each other and with one or several steel wires. If this sensor, which is connected normally to the structural element which is to be monitored over a distance of 1 m through a bearing material, is exposed to tensile load, the sensor effect is created by a loss of light intensity as a result of micro-bending. This light attenuation allows an integral measurement of the changes in the length of the sensor with a long-term accuracy of

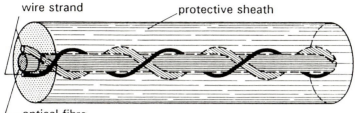

Figure 6.18 Crack detection sensor.

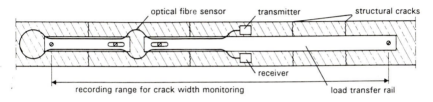

Figure 6.19 Crack width sensor.

±0.2 mm. The location of these extension changes (e.g. cracks) is carried out with an optical time domain reflector (OTDR) by reflection measurements and a local resolution of ±0.5 m (Figure 6.18).

6.5.2 The crack width sensor

The crack width sensor is an optical fibre sensor bent into a loop which is fixed to two bars. These bars are connected to the structure on the right and on the left of the joints or cracks which are to be monitored. The sensor monitors and measures permanent changes in the joints or crack width with a measuring sensitivity of 0.02 mm and a measurement range of 0.1–10 mm (Figure 6.19).

6.5.3 The reflector sensor

A light signal is passed through the optical fibre sensor. The reflectors reflect part of the light while non-reflected light travels to the next reflector. In this way, up to 30 measuring points can be monitored with a single sensor. Velocity measurements provide data on deformations.

Another design is the parallel arrangement of several optical fibre sensors. In this case, light signals are passed one after the other through each optical fibre sensor with the aid of a multiplexer. With velocity measurements, the length of the respective optical fibre sensor is measured

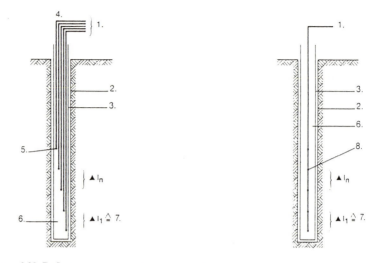

Figure 6.20 Reflector sensor: 1, electronic analysis; 2, bore hole; 3, glass fibre composite pipe; 4, optical fibre sensors; 5, reflective end of sensor; 6, synthetic resin; 7, basic length as required; 8, semi-permeable reflector.

and then compared with the sensor length of the parallel arrangement (Figure 6.20).

6.5.4   Corrosion monitoring system

Due to the increase in environmental damage (acid rain and de-icing salts), the influence of chemicals becomes bigger and bigger. For that reason, the Institut für Bauforschung (Institute for Construction Research) in Aachen have developed so-called corrosion cells [12]. In the Schiessbergstrasse bridge, a total of nine corrosion cells were integrated into the superstructure and the side walk. In the Nötsch bridge, five corrosion cells are integrated in the side walk. These cells indicate the progression of carbonation and an eventual penetration of chlorides. Here, all parameters having a decisive influence on the chemical changes of the concrete and consequently on the reinforcing steel are measured (Figure 6.21).

6.5.5   Data processing

The measured values are processed by a personal computer in a measuring chamber using specially developed software. A telephone line from the measurement chamber to the office of the client allows an inquiry and storage of the measured values. Comparing the measured values with the theoretical values, it is possible to indicate immediately considerable

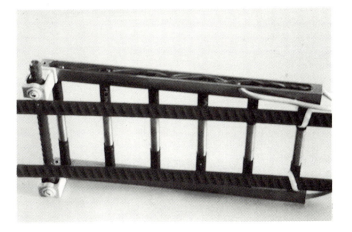

Figure 6.21 Chemical sensors for bridges.

divergences by so-called alert indications. This automatic processing and control of all data, which is important for all changes in the building structure, facilitates a significant reduction in the maintenance costs of the building. This permanent observation allows changes in the building structure to be recognised at any time and countermeasures to be started at an early stage. It is known that the costs for maintenance can be reduced considerably when it is done in good time.

6.6   Trial loadings

6.6.1   Ulenbergstrasse bridge, Düsseldorf

The first application of optical fibre sensors was the Ulenbergstrasse bridge in Düsseldorf. Both measurement systems, optical fibre sensors in the prestressing tendons and also directly embedded in the concrete, were installed. The results of this optical fibre monitoring system were stored on tape and then transferred to a computer. The attenuation curve of the monitoring results recorded in 1 week in August 1987 showed that no changes had taken place in the structure. In addition, the measurements taken in the last 5 years do not show any changes in the load-bearing structure (Figures 6.22, 6.23).

6.6.2   Marienfelde bridge, Berlin

The Marienfelde bridge in Berlin, a two-span double T-beam construction, has external prestressing with glass-fibre tendons without bond.

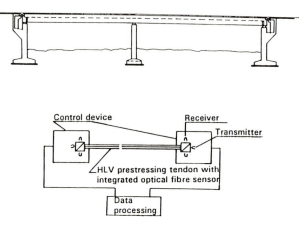

Figure 6.22 Measurement principle, Ulenbergstrasse bridge.

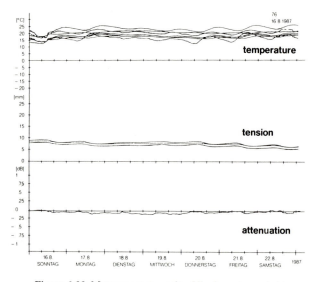

Figure 6.23 Measurement results, Ulenbergstrasse bridge.

Each of the seven glass-fibre tendons has integrated sensors. A series of optical fibre sensors has also been placed directly in the concrete or retro-mounted. The arrangement of the sensors in the bridge is shown in Figure 6.24. During a trial loading carried out in November 1989, 250 concrete slabs in five layers, with a weight of 1 metric tonne per slab, were used. The live load was applied twice in this trial loading. The cracks generated in this way (there was partial prestressing in this case) could be accurately

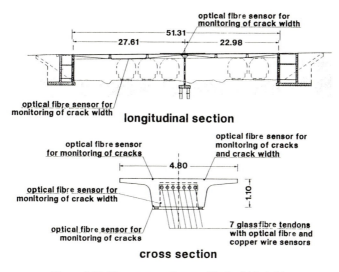

Figure 6.24 Measurement layout, Marienfelde bridge.

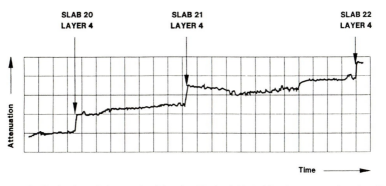

Figure 6.25 Monitoring of the test load for the Marienfelde bridge by means of optical fibre sensors embedded in the concrete.

recorded with the aid of optical fibre sensors. A crack width sensor, provided with an 8-m extension on the underside of the bridge in the long span, detects the position of the individual slabs on the surface of the deck slab (Figure 6.25).

6.6.3 Schiessbergstrasse bridge, Leverkusen

This bridge (bridge classification 60/30) is designed with limited prestressing comprising 27 glass-fibre prestressing tendons. Three glass-fibre bars per tendon are provided with sensors and there are four additional optical fibre sensors integrated directly into the concrete on the upper and four on the lower side of the slab. A trial loading of the Schiessberg-strasse bridge was carried out on 31st of March 1992, using two trucks with a total load of 50 tonnes, illustrating the efficiency of the optical fibre sensors embedded in the concrete construction. By connecting two highly sensitive sensors for this trial load, it was possible to measure additional elongations in the micrometre-sector. These high sensitivity sensors are normally only used for structural monitoring in special cases, and were particularly applied for demonstration purposes in this instance.

The sensors, which are now permanently connected in the bridge, only register additional extensions in the magnitude of 0.2 mm, thus enabling the crack behaviour of the concrete to be monitored.

The bridge has a highly sophisticated permanent monitoring system with the possibility of on-line diagnosis. The main benefits are:

- prestressing with intelligent glass-fibre tendons monitored by optical fibres;
- crack control of the concrete by optical fibre sensors embedded directly in the concrete;
- integrated chemical sensors in the concrete for measuring the carbonation depth and chloride penetration;
- all data able to be checked from a personal computer via a telephone line from the bridge to the client's office (Figures 6.26, 6.27).

6.6.4   Nötsch bridge, Kärnten, Austria

The Nötsch bridge also has a highly sophisticated permanent monitoring system with the possibility of on-line diagnosis with the aid of a telephone line to transfer the data directly to the client's office. Similar to the Schiessbergstrasse bridge, the suitability of the sensors has been proved by a trial loading. It was carried out in May 1992 using two trucks with a weight of 22 metric tonnes each. As expected, the integral additional elongation of approximately 50 μm in the second span corresponded to a deflection of approximately 2–3 mm (Figures 6.28, 6.29).

6.7   Conclusion

The first applications, in particular the bridges of the Ulenbergstrasse, Marienfelde, Schiessbergstrasse and Nötsch, are only the beginning of a whole series of other possible applications. On the 7th of December 1992, the Institut für Bautechnik in Berlin granted general approval for the glass-fibre composite prestressing system, which is now the first pre-

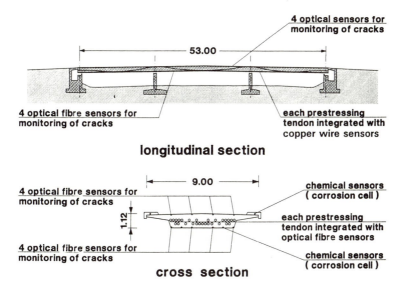

Figure 6.26 Sensor layout, Schiessbergstrasse bridge.

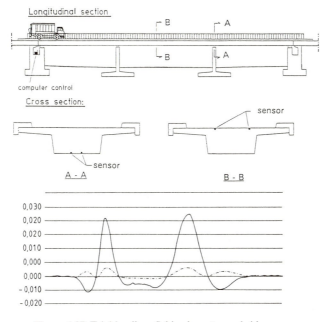

Figure 6.27 Trial loading, Schiessbergstrasse bridge.

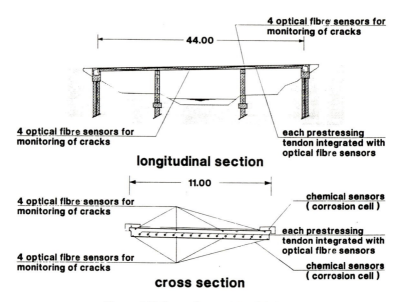

Figure 6.28 Sensor layout, Nötsch bridge.

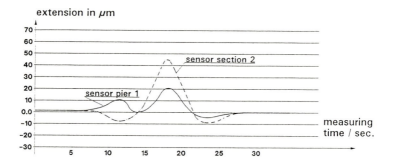

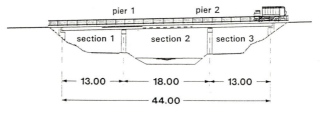

Figure 6.29 Trial loading, Nötsch bridge.

INSTITUT FÜR BAUTECHNIK

Anstalt des öffentlichen Rechts

1000 Berlin 30, 7. Dezember 1992
Reichpietschufer 74-76
Telefon: (030) 264 87-266
Teletex: 308258
Telefax: (030) 264 87-320
GeschZ.: I 1-1.13.1-67

ZULASSUNGSBESCHEID

Der

Zulassungsgegenstand: Spannverfahren HLV

wird hiermit allgemein bauaufsichtlich/baurechtlich zugelassen.

Antragsteller: Strabag Bau-AG   SICOM, Gesellschaft
                          Siegburger Str. 241   für Sensor- und Vor-
                          5000 Köln 21   spanntechnik mbH
                                                  Gremberger Str. 151a
                                                  5000 Köln 91

                          Bayer AG
                          5090 Leverkusen

Geltungsdauer bis: 30. November 1997

Zulassungsnummer: Z-13.1-67

Der zugelassene Gegenstand darf nur verwendet werden, wenn seine Herstellung
überwacht ist und dies am Verwendungsort geprüft werden kann.

Dieser Zulassungsbescheid umfaßt 23 Seiten und 12 Anlagen.

Figure 6.30 General approval for the glass-fibre composite prestressing system, granted by the Institut für Bautechnik, Berlin, December 1992.

stressing system worldwide based on fibre composite material. The scope of application includes all types and degrees of prestressing, with or without bond, temporary or permanent soil or rock anchors, as well as tension cables for transmission masts or bracings of any kind. Work is continuing in the development of fibre composite materials into intelligent high-tech products with tensile strength and media resistance. The possibility of monitoring the integrity of prestressed concrete structures and earth anchor systems is also being researched. This development will give prestressed concrete design a decisive impulse in the future, as concrete structures could have integrated sensitive nerve fibres just like organic bodies and be more durable. Any change in their stress–strain behaviour can be immediately detected at any location within the structure. Comprehensive control of concrete structures will considerably reduce the maintenance costs in the long run, because even minimal changes in the load-bearing behaviour are recorded and the necessary repair and reconstruction work can be undertaken at a very early stage.

References

1. Miesseler, H.-J. and Preis, L., Heavy-duty composite bars made of glass fibre as reinforcement in concrete and foundation constructions, *Bauen mit Kunststoffen*, **2** (1988) 4–14.
2. Rehm, G. and Franke, L., Synthetic resin bonded glass fibre bars as reinforcement in concrete construction, *Die Bautechnik* **4** (1974) 115–120.
3. Rehm, G., Franke, L. and Patzak, M., *Investigations Regarding the Question of Introduction of Forces in the Synthetic Resin Bonded Glass Fibre Bars. Issue 304. The German Committee for Reinforced Concrete—Berlin*, Wilhelm Ernst, 1979.
4. Faoro, M., The loadbearing behaviour of resin bonded glass fibre tendons in the field of end-anchoring and fractures in concrete, University of Stuttgart, Institute for Materials in the Field of Civil Engineering (Institut für Werkstoffe im Bauwesen), Stuttgart, dissertation, 1988, IWB-news 1988/1.
5. Waaser, E. and Wolff, R., A new material for prestressed concrete, HLV—heavy duty composite bars comprising glass fibres, *Beton* **36** (1986) 245–250.
6. Vollrath, F. and Miesseler, H.-J., Glass fibre prestressing for concrete bridges—experience gained during construction of the Ulenbergstrasse Bridge, *TIEFBAU BG* **4** (1987) 206–211.
7. Franz, A. and Miesseler, H.-J., Bridge at Berlin-Marienfelde: a research project. External prestressing and permanent monitoring by means of integrated sensors, in *German Concrete Congress*, Hamburg, 1989.
8. Wolff, R. and Miesseler, H.-J., Application and experience with intelligent prestressing systems based on fibre composite materials, in *XI FIP-Congress*, Hamburg, 1990, pp. R63–R67.
9. Miesseler, H.-J. and Levacher, K., Monitoring stressing behaviour with integrated optical sensors, in *13th IABSE Congress*, Helsinki, 1988 (in German).
10. Miesseler, H.-J. and Lessing, R., Monitoring of load bearing structures with optical fibre sensors, in *IABSE*, Lisbon, 1989.
11. Miesseler, H.-J. and Wolff, R., Experience with the monitoring of structures using optical fibre sensors, in *XI FIP Congress*, Hamburg, 1990, pp. Q12–Q17.
12. Schiessl, P. and Raupach, M., Chloride-induced corrosion of steel in concrete—investigations with a concrete corrosion cell, Institute for Building Research, Technical University of Aachen, in *The Life of Structures*, Brighton, UK, 1989.

7 Strengthening of structures with advanced composites

U. MEIER, M. DEURING, H. MEIER
and G. SCHWEGLER

7.1 Introduction

7.1.1 Formulation of the problem

Often it becomes necessary to strengthen existing structures or parts of them. The reasons that make this sort of reinforcement necessary can be summarized as follows. First, a change in the use of a structure may produce internal forces in individual structural parts that exceed the existing cross-sectional strengths. These increased internal forces may be a result of higher loading or less favorable configuration of existing loading. Structural components may also need reinforcement because external influences have caused damage that has reduced the cross-sectional resistance. The object of repairing such damage is to restore the original cross-sectional strength. Another possibility is misdesign of a structure or parts of it. This includes all cases where the cross-sectional strength at crucial points is too low, so that either the cross-sectional safety or the overall safety of the respective structure or structural element fails to comply with existing codes. Finally, poor construction workmanship may mean that the cross-sectional strengths originally calculated are not achieved. For instance, the as-built cross-sectional dimensions may be considerably smaller than those planned. Or it can happen that individual rebars are incorrectly set, interchanged or even missing, which reduces the cross-sectional strength substantially.

There are a number of ways to strengthen a given structure or parts of one, depending on the type of construction and the given situation. Bonded external steel reinforcement is one possible way of achieving such structural strengthening. This method was originally invented in France [1] in the mid-1960s; in the early 1970s it was further developed in Switzerland [2, 3] and England [4] and is nowadays state-of-the-art in Western Europe.

7.1.2 Reasons for replacing steel by advanced composites

In previous papers [5–8], the advantages and disadvantages of post-strengthening by means of steel plates have been discussed, as well as the reasons for replacing steel plates with advanced composites, especially with carbon-fibre reinforced plastics (CFRPs), above all in bridge construction.

1. The most important reason is clear from the following observation. Within the framework of long-term creep tests on beams strengthened with steel plates at the EMPA, the residual strength was determined after 15 years of exposure to weathering without de-icing salts. In several cases, spots of corrosion (diameter 1–5 mm) were discovered at the joint between the steel and the adhesive. These spots are growing and the strengthening system will finally fail due to corrosion.

2. Steel plates are heavy. Therefore their handling on construction sites, e.g. inside a box girder, is not convenient. Often expensive scaffolding is required to bond the steel plates to the structure.

(a)

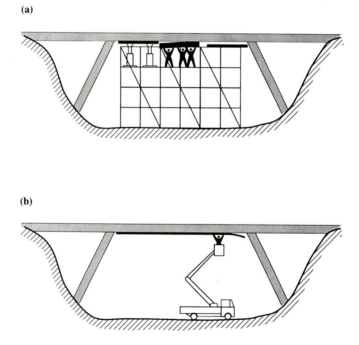

(b)

Figure 7.1 (a) Steel plate bonding. This classical strengthening method is today state-of-the-art in Western Europe. Advantages: low material cost, common use. Disadvantages: corrosion problems, due to heavy plates, difficult handling on construction site and high labour and scaffolding cost, limited length of steel plates, therefore difficult joint problems. (b) CFRP sheet bonding. New strengthening method. Advantages: no corrosion problems, very light weight therefore easy handling on construction site, low labour and no scaffolding cost, no joints necessary because endless sheets available, outstanding fatigue performance. Disadvantages: high materials cost, not yet in common use.

3. Due to the weight of the steel plates their length is in general restricted to 6, or a maximum 10 m. If a greater length is required joints are necessary. Such joints cannot be welded since this would destroy the adhesive bonding.
4. If bonded steel plates are compression loaded, they may fall off in a buckling mode.

Advanced fibrous composites offer the designer in the construction industry an outstanding combination of properties not available from other materials. Fibres such as glass, aramid and carbon can be introduced in a certain position, volume fraction and direction in the matrix (e.g. epoxy resin) to obtain maximum efficiency. Other advantages offered by advanced composites are lightness, corrosion resistance, for some also an outstanding fatigue performance, and greater efficiency in construction compared with the more conventional materials. Advanced composite sheets can replace steel plates with overall cost savings emanating from the simplicity of the strengthening method because:

1. they do not corrode;
2. they are easy to handle on the construction site and can be bonded to the structure with a scissors-lift or similar lift without expensive scaffolding (see Figure 7.1);
3. they are available endless on bobbins, therefore no joints are necessary;
4. if some of them (e.g. carbon fibre reinforced epoxy resins) are subjected to compressive stresses they do not fall off [9];
5. some (e.g. carbon-fibre reinforced epoxy resins) show an out-standing fatigue behaviour.

7.1.3 Optimal fibre for the strengthening task

The most suitable manufacturing process for advanced composites is pultrusion. This process is described in Chapter 1, Section 1.8.2. A typical cross-section through such an advanced composite is shown in Figure 7.2. As mentioned earlier, glass, aramide and carbon fibres are the prime candidates for the strengthening applications. When we started R&D work in this field in the early 1980s, we decided to use carbon fibre. This decision is based on the comparisons given in Table 7.1.

7.1.4 Scope of the R&D work at the EMPA

Since 1984, in static and fatigue loading tests at the EMPA laboratories, carbon-fibre reinforced plastic (CFRP) sheets have successfully been em-ployed for the post-strengthening of flexural beams with a span up to 7 m. The results of this comprehensive research programme show that the

Figure 7.2 Scanning electron micrograph (SEM) of a carbon-fibre reinforced plastic (CFRP) sheet described in Table 7.2. The diameter of a single filament is about $7\,\mu m$.

Table 7.1 Comparison of important characteristics of fibre reinforced plastic sheets produced from different fibres; the fibre volume fraction for such sheets is typically about 65%; the filaments are parallel and unidirectional; this comparison applies specifically to the sheet bonding technique for rehabilitation of concrete structures

Characteristics	Sheets with fibres of		
	E-glass	Aramide	HT carbon
Tensile strength	Very good	Very good	Very good
Compressive strength	Good	Poor	Good
Stiffness	Poor	Good	Very good
Static fatigue	Poor	Good	Excellent
Cyclic fatigue	Fair	Good	Excellent
Density	Fair	Excellent	Good
Alkaline resistance	Poor	Good	Very good
Cost	Good	Fair	Sufficient

calculation of flexure in reinforced concrete elements post-strengthened with untensioned or with tensioned carbon-fibre reinforced epoxy-resin plates can be performed analogous to conventional reinforced or pre-stressed concrete. In both cases, long-time fatigue tests on 6-m girders illustrated the outstanding fatigue performance of this strengthening technique. A new approach also allows an increase in shear strength using post-strengthening systems. In 1991, for the first time real structures were strengthened with CFRP in Switzerland, e.g. the multispan box beam Ibach bridge near Lucerne with a total length of 228 m. The City Hall of Gossau St. Gall followed in the same year. In addition, the historic wooden bridge near Sins was strengthened for heavy trucks.

7.2   Strengthening with non-pretensioned CFRP sheets

In the years 1984–1989, carbon-fibre reinforced epoxy-resin was success-fully employed for the first time at the EMPA for post-strengthening purposes. Loading tests were performed on 26 flexure beams having a span of 2 m and 1 beam having a span of 7 m [5]. The research work shows the validity of the strain-compatibility method in the analysis of cross-sections. This implies that the calculation of flexure in reinforced concrete elements post-strengthened with carbon-fibre reinforced epoxy-resin sheets can be performed analogous to conventional reinforced con-crete elements. The work also shows that possible occurrence of shear cracks as shown in Figure 7.3 may lead to a peeling-off of the strength-ening sheet. Thus, the shear crack development represents a design criterion. Flexural cracks are spanned by the CFRP sheet and do not in-fluence the loading capacity. In comparison to the unstrengthened beams, the strengthening sheets lead to a much finer cracking distribution [7] as demonstrated in Figure 7.4. A calculation model [5], developed for the CFRP sheet anchoring agrees well with experimental results.

Differences in the coefficients of thermal expansion of concrete and the carbon-fibre reinforced epoxy-resin result in stresses at the joints with

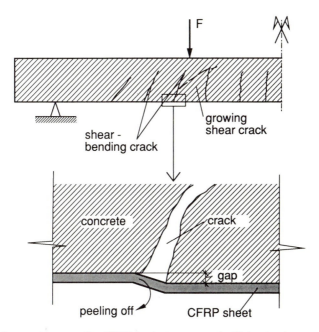

Figure 7.3 In certain cases, the CFRP laminate may peel off due to shear cracks. This was especially observed near the loading points in beams with thick laminates and high reinforcement.

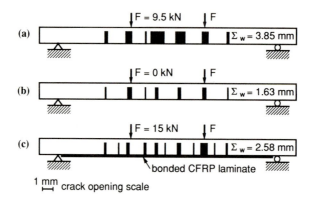

Figure 7.4 Crack development. (a) Beam of 2 m span without post-strengthening, load $2F = 19$ kN, total crack widths 3.85 mm. (b) Same beam after unloading, residual total crack width 1.63 mm. (c) Same beam after strengthening with a 0.75-mm thick CFRP sheet, increased load $2F = 30$ kN, total crack widths 2.58 mm.

changes in temperature. After 100 frost cycles ranging from $+20$ to $-25°C$, no negative influence on the loading capacity of the three post-strengthened beams was found [5].

The following failure modes were observed in the load tests:

1. Tensile failure of the CFRP sheet (see Figure 7.5(1)). The sheets failed more or less suddenly, with a sharp explosive snap. The impending failure was always announced far in advance by cracking sounds.
2. Classical concrete failure in the compressive zone of the beam (Figure 7.5(2)).
3. Continuous peeling-off of the CFRP sheets due to an uneven concrete surface (Figure 7.5(3)). For thin sheets (less than 1 mm) applied with a vacuum bag, an extremely even bonding surface is required. If the surface is too uneven, the sheet will slowly peel off during loading.
4. Shearing of the concrete in the tensile zone (Figure 7.5(3)) (also observed as secondary failure).
5. Interlaminar shear (Figure 7.5(4)) within the CFRP sheet (observed as secondary failure).
6. Failure of the reinforcing steel in the tensile zone (Figure 7.5(5)). This failure mode was only observed during fatigue tests.

The following failure modes have not yet been observed but are theoretically possible:

1. Cohesive failure within the adhesive (Figure 7.5(6)).
2. Adhesive failure at the interface CFRP sheet/adhesive (Figure 7.5(7)).
3. Adhesive failure at the interface concrete/adhesive (Figure 7.5(8)).

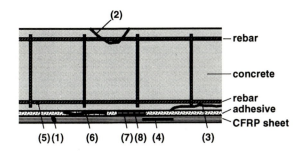

Figure 7.5 Failure modes: (1) tensile failure of the CFRP sheet; (2) concrete failure in the compressive zone; (3) continuous peeling off of the CFRP sheet due to an uneven concrete surface; (4) interlaminar shear within the CFRP sheet; (5) failure of the steel rebars; (6) cohesive failure within the adhesive; (7) adhesive failure at the CFRP sheet/adhesive interface; (8) adhesive failure at the concrete/adhesive interface.

For post-strengthening with CFRP sheets, we recommend the design rule that the CFRP sheets should fail during yielding of the steel reinforcing bars before failure of the concrete in the compressive zone. Yielding of the steel bars should not occur before reaching the permitted loads.

Kaiser [5] investigated at the EMPA a beam with a 2.0 m span under fatigue loading. The cross-section was 300 mm wide and 250 mm deep. The existing steel reinforcement consisted of two rebars of 8 mm diameter in the tension as well as the compression zones. This beam was post-strengthened with a glass/carbon-hybrid sheet with the dimensions 0.3 × 200 mm² (see Table 7.2, sheet type no. 1). The fatigue loading was sinusoidal at a frequency of 4 Hz. The test set-up corresponded to a four-point flexure test with loading at the one-third points. The calculated stresses in the hybrid sheet and the steel reinforcement are listed in Table 7.3. After 480 000 cycles, the first fatigue failure occurred in one of the two rods in the tension zone. After 560 000 cycles, the second rod broke at another cross-section. After 610 000 cycles, a further break was observed in the first rod and after 720 000 cycles, in the second rod. The first damage appeared after 750 000 cycles. It was in the form of fractures of individual rovings of the sheet. The beam exhibited gaping cracks, which, however, were bridged by the hybrid sheet. The relatively sharp concrete edges rubbed against the hybrid sheet at every cycle. After 805 000 cycles, the sheet finally failed. This test was executed with unrealistically high steel stresses. The goal of the test, however, was to gain insight into the failure mechanism after a complete breaking of the steel reinforcement. This goal was achieved. It was remarkable to observe how much the hybrid sheet could withstand after failure of the steel reinforcement.

In 1991–1992 [9], the EMPA performed a further fatigue test on a beam with a span of 6.0 m under more realistic conditions. The dimen-

Table 7.2 Properties of hybrid and CFRP sheets

Property	Sheet type no.				
	1	2	3	4	5
Fibre type	2/3 T 300 1/3 E-glass	T 300	–	T 700	M 46 J
Fibre volume fraction (%)	–	70	51	66	70
Longitudinal strength (MPa)	960	2000	1900	2300	2600
Longitudinal Young's modulus (GPa)	80	147.5	129	152	305
Strain at failure (%)	1.15	1.36	1.47	1.51	0.85
Density (g/cm^3)	–	1.58	1.46	1.45	–

Table 7.3 Fatigue loading and stresses of the 2-m beam

Loads (kN)	Stresses (MPa)	
	Rebars	Hybrid sheet
Minimum 1	21	11
Maximum 19	407	205

sions and reinforcement of the beam can be seen in Figure 7.6. The total carrying capacity F of this beam amounted to 610 kN without the CFRP. Through the bonding of a CFRP sheet of dimensions $200 \times 1\,mm^2$ (sheet type no. 2, Table 7.2), the carrying capacity was increased by 32% to 815 kN. In Figure 7.7, typical load–deflection diagrams are presented for this type of beam. The calculated stresses in the CFRP sheet and the steel reinforcement are given in Table 7.4. The beam was subjected to this loading for 10.7 million cycles. The crack development was observed after every 2 million cycles [9]. After 10.7 million cycles, the tests were continued in a climatic room. The temperature was raised from room temperature to 40°C and the relative humidity to at least 95%. The goal of this test phase was to verify that the bonded CFRP sheet can withstand very high humidity with simultaneous fatigue loading. At the beginning of this test phase, the CFRP sheet was already nearly completely saturated with water. After a total of 12.0 million cycles, the first steel reinforcement failed due to fretting fatigue. The joint between the CFRP sheet and the concrete did not present even the slightest of problems. In the continuation of the test, the external loads were held constant (Table 7.4) whereas the stresses in the reinforcement steel and the CFRP sheet correspondingly increased. After 14.09 million cycles, the second reinforcement steel failed, likewise due to fretting fatigue. The cracks bridged by the CFRP sheet rapidly grew. After failure of the third reinforcement

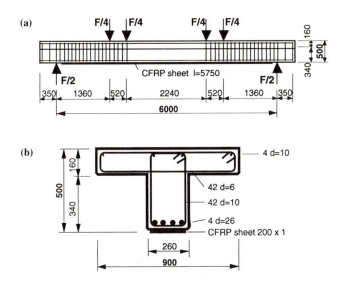

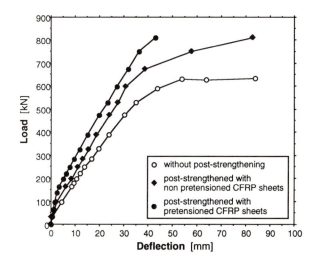

Figure 7.6 (a) View and (b) cross-section of the beams with 6-m span (all dimensions are in millimetres).

Figure 7.7 Typical load–deflection curves for beams shown in Figure 7.6.

Table 7.4 Fatigue loading and stresses of the 6-m beam

Loads (kN)	Stresses (MPa)	
	Rebars	CFRP sheet
Minimum 125.8	131	102
Maximum 283.4	262	210

rod due to yielding of the remaining overloaded steel, the CFRP sheet was sheared off.

7.3   Strengthening with pretensioned CFRP sheets and shear strengthening

A further fatigue test was carried out at the EMPA, analogous to that described in the previous section. The only difference was that the CFRP sheet was prestressed (50% of strength of sheet type no. 2, Table 7.2). Thirty million cycles were performed without any evidence of damage.

In recent months, a new method has been developed and tried out in several tests. This allows an effective strengthening of the shearing force areas without the use of steel. One possible implementation is illustrated in the beam cross-section of Figure 7.8. The inner stirrup reinforcement is supplemented by a stressed or limply applied external strengthening made of advanced composite materials. These are braided or unidirectional in form. Depending on the application, carbon fibres may be employed as well as aramid and glass fibres. The pretensioning material is wrapped around the cross-section on one side and anchored on the opposite side in the compression zone. The modulus of elasticity and geometry should be chosen in order to minimise the loss of tensioning force due to creep of the element to be strengthened and relaxation of the prestressing material.

Many bridges in need of rehabilitation are deficient not only with respect to flexural resistance. Often it is necessary to strengthen mainly the shear resistance. In order to contribute at once to the strengthening and to relieve the inner stirrup reinforcement, the reinforcement located externally to the element should be additionally prestressed. In this way, crack formation in the shear force region can be precluded or the cracks more finely distributed, in case shear cracks develop.

Another method of strengthening reinforced concrete structures in shear by bonding externally fibre-reinforced plastic composite sheets to the lateral concrete surfaces through epoxy resins is presented by Berset [10]. Analytical models are developed describing the design variables

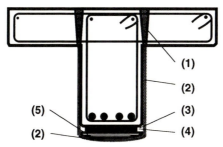

Figure 7.8 Cross-section of beam with shear strengthening arrangement: (1) anchorage zone; (2) shear strengthening element of advanced composites; (3) adhesive; (4) CFRP sheet for flexural strengthening; (5) CFRP plate for load distribution.

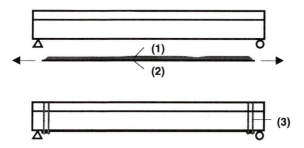

Figure 7.9 Procedure for applying a prestressed sheet: (1) adhesive; (2) prestressed CFRP sheet; (3) shear strengthening arrangement according to Figure 7.8.

and the ultimate achievable force involving the maximum applied shear stresses at the interface between the composite and the concrete, so that debonding of the system does not occur. Analytical solutions describing the collapse mechanisms are derived. An experimental programme was carried out confirming the analytical results and showing that the shear capacity of beams strengthened with this technique is notably improved when compared to members without composite reinforcement.

In some cases, it can be advantageous to provide additional prestressing to the flexure-strengthening sheets. In this way, the serviceability of the structure can be improved and the shearing off of the sheets due to shear failure of the concrete in the tension zone can be avoided. An initial publication of this EMPA and MIT research effort appeared recently [11]. Detailed results will soon follow [9]. The procedure for applying a prestressed sheet is shown schematically in Figure 7.9. When the pretensioning force is too high, failure of the beam due to pretension release will occur at the two ends because of the development of high shear stresses in the concrete layer just above the CFRP sheet. Therefore, the design and construction of the end regions require careful attention. Tests and calculations have shown that without special end anchoring, CFRP sheets shear off from the end zones immediately with a prestress of only 5% of their failure strength. In order to achieve a technically and economically rational prestress, considerably higher degrees of prestressing in the range of 50% are necessary.

At the EMPA, end anchorings for flexure beams in accordance with Figures 7.8 and 7.9 were developed and successfully tested. In contrast to pure shear strengthening, the advanced composites which wrap around the sheet must most definitely be prestressed. This will build up as high a multi-axial stress condition in the concrete as possible and also interlock cracks. In this way, failure at the two ends of the CFRP sheets can be avoided.

7.4 Applications

7.4.1 The Kattenbusch bridge in Germany

To the best of our knowledge, the Kattenbusch bridge in Germany is the first place in the world where fibre reinforced plastic plates were used to strengthen a bridge.

After World War II, numerous prestressed concrete bridges for motor vehicles were built in Germany employing the method of *in situ* spanwise construction [12]. These continuous multi-span bridges are mostly designed as box girders. The working joints are at the points of contraflexure where usually all of the tendons are coupled. Many of the bridges now exhibit cracks at the working joints. Usually, the bottom slab of the box girder is transversely cracked at the joint. This relatively wide crack grows into the webs with diminishing width. Thereby, it crosses the lower tendons and couplings. The main cause of these cracks is a temperature restraint which was not taken into account during previous designs [13]. In combination with other stresses, tensile stresses at the bottom increase and exceed the concrete tensile strength at the joint. As the reinforcement ratio of the bottom slab was often low, yielding of the steel occurred and wide cracks formed. Due to increased fatigue stresses, the durability of the reinforcement and the tendons was no longer assured. Thus the necessity for repair arose. In the late 1970s, Rostasy and his co-workers [14–16] developed a technique to strengthen such joints with bonded steel plates. The first successful application was the Sterbecke bridge near Hagen (Germany) in 1980. In 1986–1987 this method was used for the first time with glass-fibre reinforced plastic plates on the Kattenbusch bridge.

The Kattenbusch bridge is designed as a continuous, multi-span box girder with a total length of 478 m. It consists of 9 spans of 45 m and 2 side spans of 36.5 m each. There are 10 working joints. The depth of the twin box girder is 2.70 m. The bottom slab of the girder is 8.50 m wide. One working joint was strengthened with 20 glass-fibre reinforced plates. Each plate is 3200 mm long, 150 mm wide and 30 mm thick. Loading tests performed by Rostasy and co-workers [17] showed a reduction in the crack width of 50% and a decrease in the stress amplitude due to fatigue of 36%. The static and the fatigue behaviour were at least equal to those when using the steel plate bonding technique. From the corrosion point of view, the expectations of the glass-fibre reinforced plastic plates are much higher.

7.4.2   The Ibach bridge near Lucerne in Switzerland

The bridge to be repaired (Figure 7.10), located in the County of Lucerne, was completed in 1969. It is designed as a continuous, multi-span box beam with a total length of 228 m. The damaged span of the bridge has a length of 39 m. The box section is 16 m wide, with a central, longitudinal web.

Core borings were performed to mount new traffic signals. In the process, a prestressing tendon in the outer web was accidentally damaged, with several of its wires completely severed by means of an oxygen lance. As a result, the granting of authorisations for special, heavy loads across the bridge was suspended until after completion of the repair work. Since

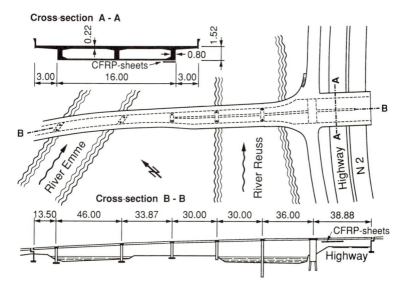

Figure 7.10 Ibach bridge near Lucerne, Switzerland (all dimensions are in metres).

the damaged span crosses National Highway N2, the traffic lanes in the direction of Lucerne on this highway had to be closed during the repair work. The work could therefore only be conducted at night.

Carbon-fibre reinforced plastics (CFRPs) are forty to fifty times more expensive per kilogram than the steel used to date (Fe 360) for the reinforcement of existing structures. Do the unquestionably superior properties of CFRPs justify their high price? When one considers that, for the repair of the Ibach bridge, 175 kg of steel could be replaced by a mere 6.2 kg of CFRP, the high prices no longer seem so outrageous. Furthermore, all the work could be carried out from a mobile platform, thus eliminating the need for expensive scaffolding.

The bridge was repaired with three CFRP sheets of dimensions 150 × 5000 × 1.75 mm^3 (2 sheets) and 150 × 5000 × 2.00 mm^3 (1 sheet) according to sheet type no. 3 in Table 7.2. A loading test with an 84-tonne vehicle demonstrated that rehabilitation work with the CFRP sheets was very satisfactory. The experts participating in the repair of the Ibach bridge were pleasantly surprised at the simplicity of applying the 2-mm thick and 150-mm wide CFRP sheets.

7.4.3   The historic wooden bridge near Sins in Switzerland

The covered wooden bridge near Sins in Switzerland (Figure 7.11) was built in 1807 in accordance with the design of Josef Ritter of Lucerne. On the Sins side, the original supporting structure is almost completely preserved, even today. The Chams side was blown up for strategic purposes on November 10, 1847 during the Civil war. In 1852, the destroyed half of the bridge was rebuilt with a modified supporting structure. On

Figure 7.11 Historic wooden bridge near Sins, Switzerland with two spans of 30.8 m each.

the Sins side, the supporting structure consists of arches. These are strengthened with suspended and trussed members. On the Chams side, the supporting structure is made up of a combination of suspended and trussed members with interlocking tensioning transoms. Originally the bridge was designed for horse-drawn vehicles. Today, vehicles with a load of 20 tonnes are permitted.

In the 186-year history of the bridge, a great variety of rehabilitation efforts have been undertaken. Loading tests performed by the EMPA and the ETH Zürich indicated that the pavement and several cross-beams no longer met the requirements of heavy traffic. A project involving the construction of a pretensioned concrete bridge several hundred metres upstream was opposed by the residents. Thus, in 1992 the wooden bridge urgently had to be rehabilitated. It was decided to replace the old wooden pavement with 20 cm thick bonded wooden planks, transversely pretensioned. This technique described in the Ontario Bridge Design Code in 1983 was further developed at the ETH Zürich. Two of the most highly loaded cross-beams were strengthened by the EMPA using carbon fibre reinforced epoxy resin sheets. Each of these cross-beams was constructed of two solid oak beams placed one upon the other. A cross-section of the bridge with the strengthening is shown in Figure 7.12. In order to increase the depth, wooden blocks have originally been inserted between the beams. The lower beams were 37 cm deep and 30 cm wide, the upper beams 30 cm deep and 30 cm wide.

Cross-beam no. 14 was strengthened with 1.0-mm thick CFRP sheets made of high modulus fibres (sheet type no. 5, Table 7.2); on the upper side, the width amounted to 250 mm and on the lower side, 200 mm. Cross-beam no. 15 was strengthened with 1.0-mm thick CFRP sheets made of high strength fibres (sheet type no. 4, Table 7.2); on the upper side, the width amounted to 300 mm and on the lower side, 200 mm.

Before bonding the sheets, the bonding surface was planed with a portable system (Figure 7.13). The bonding and installation were accomplished with the same material and a similar clamping system (Figure 7.14) as for the Ibach bridge [18].

The strengthened cross-beams of the Sins bridge, subjected to extremely high loading and reinforced with CFRP sheets, have provided practical experience and confidence in this method for preservation of historic bridges. Thus, in the future, similar structures may be rehabilitated in this manner.

The historic wooden bridge near Sins is a valuable structure, both from aesthetic and technical viewpoints. It is also of value historically and under protection as a national monument. For the post-strengthening of such structures, the technique with CRP plates is especially suited since the thin but extremely stiff and strong plates are hardly noticed and therefore do not detract from the original design of the structure.

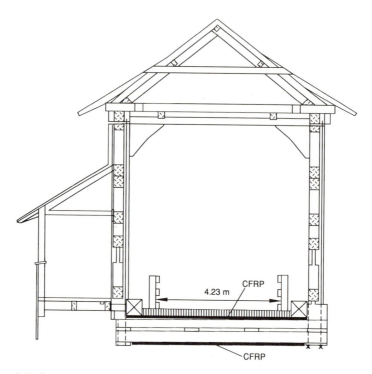

Figure 7.12 Cross-section of the historic bridge near Sins. Selected cross-beams were strengthened with CFRP sheets.

Figure 7.13 Portable system to plane the surface of the wooden cross-beams.

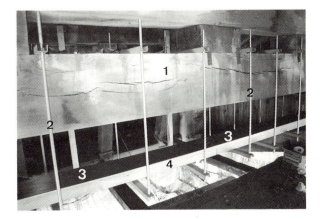

Figure 7.14 CFRP sheet (1 mm) ready to be bonded to the wooden cross-beam: (1) wooden cross-beam; (2) rods to press the CFRP sheet to the wooden beam; (3) CFRP sheet still without adhesive; (4) plate to press the CFRP sheet to the wooden beam.

7.4.4   The City Hall of Gossau St. Gall in Switzerland

The City Hall of Gossau St. Gall was the first building in Switzerland where the CFRP sheet strengthening technique was used. Within a renovation programme it was decided that an elevator should be added. As a result, a rectangular hole had to be cut in a concrete slab. Before cutting, the future edges of the hole were strengthened with CFRP sheets as shown in Figure 7.15. Due to aesthetic reasons the architects were persuaded to use thin high strength CFRP sheets instead of thick steel plates. After painting the CFRP sheets 'disappear'.

7.5   Outlook

In the expanding range of non-metallic tensile elements based on high strength fibres, CFRP sheets, the non-corrosive tensile elements composed of endless parallel carbon filaments in an epoxy resin, may be considered to be a very promising alternative to steel plates for strengthening applications where long-term durability is required. Carbon sheets combine the qualities of very high strength with an outstanding fatigue performance and light weight for easy handling. The sheets are durable under practically every type of environmental attack which may occur in or around concrete structures.

In Switzerland, the application of non-tensioned CFRP sheets to strengthen existing structures successfully crossed the threshold to ap-

Figure 7.15 Concrete slab of the City Hall of Gossau strengthened with black CFRP sheets. The part of the slab which will be cut in is indicated by a dashed line. The work was executed by Stahlton AG, 8034 Zürich, Switzerland.

plication in 1991. For wider acceptance of this efficient strengthening method, European design recommendations and codes are required.

Bending and shear strengthening methods with pretensioned CFRP sheets are not as developed. It will take at least another 2 years of R&D work to cross the threshold to applications since the handling of this pretensioning method is not yet practical.

Today, the use of CFRP sheets in civil engineering is in its infancy, but there are clear indications that it will be an excellent choice for a multitude of rehabilitation projects on bridges, dams, tubes, high pressure pipes, buildings or even historic monuments.

References

1. L'Hermite, R. and Bresson, J., Béton armé par collage d'armature, in *Colloque RILEM*, UIT, Paris, 1967, ed. Eyrolles, 1971, Vol. II, p. 175.
2. Ladner, M. and Flüeler, P., Versuche an Stahlbetonbauteilen mit geklebter Armierung, *Schweizerische Bauzeitung* **92** (1974) 463.
3. Ladner, M. and Weder, Ch., EMPA-Report No. 206, Concrete structures with bonded external reinforcement, 1981, EMPA Dübendorf, CH 8600 Dübendorf, Switzerland.
4. James, R. and Swamy, R.N., Composite behaviour of concrete beams with epoxy bonded external reinforcement, *Int. J. Cement Composites* **2** (1980) 91.
5. Kaiser, H.P., Strengthening of reinforced concrete with epoxy-bonded carbon-fibre plastics, Doctoral Thesis, Diss. ETH No. 8918, 1989, ETH Zürich, CH-8092 Zürich, Switzerland (in German).
6. Meier, U., Bridge repair with high performance composite materials, *Mater. Techn.* **15** (1987) 225–228 (in German and in French).

7. Meier, U. and Kaiser H.P., Strengthening of structures with CFRP laminates, in: *Proc. Advanced Composite Materials in Civil Engineering Structures*, MT Div, ASCE, Las Vegas, 1991.

8. Meier, U., Carbon fibre-reinforced polymers: modern materials in bridge engineering, *Struct. Eng. Int.* **2** (1992) 7–12.

9. Deuring, M., Post-strengthening of concrete structures with pretensioned advanced composites, EMPA Research Report No. 224, EMPA Duebendorf, CH-8600 Dueben-dorf/Switzerland, 1993.

10. Berset, J.-D., Strengthening of reinforced concrete beams for shear using FRP composites, M.S. Thesis, Department of Civil Engineering, Massachusetts Institute of Technology, Cambridge, MA, 1992.

11. Triantafillou, T.C., Deskovic, N. and Deuring, M., Strengthening of concrete structures with prestressed fibre reinforced plastic sheets, *ACI Struct. J.* **89** (1992) 235–244.

12. Rostasy, F.S. and Ranisch, E.-H., Strengthening of bridges with epoxy bonded steel plate, *Symp. Int. Assoc. for Bridge and Structural Engineering (IABSE) on Maintenance, Repair and Rehabilitation of Bridges*, 1982, pp. 112–122.

13. Ivanyi, G. and Kordina, K., Defects in the region of construction joints of prestressed concrete bridges, German Technical Contribution, in *8th FIP Congress*, London, 1978.

14. Rostasy, F.S., Ranisch, E.H. and Alda, W., Strengthening of prestressed concrete bridges in the region of working joints with coupled tendons by bonded steel plates, Part 1, Forschung, Strassenbau und Strassenbautechnik, Heft 326, Bonn 1980 (in German).

15. Rostasy, F.S. and Ranisch, E.H., Strengthening of prestressed concrete bridges in the region of working joints with coupled tendons by bonded steel plates, Part 2, BMV-Forschungsbericht no. 15,099, Bonn, Braunschweig, 1981 (in German).

16. Rostasy, F.S. and Ranisch E.H., Strengthening of reinforced concrete structural members by means of bonded reinforcement, *Betonwerk- und Fertigteiltechnik*, 1981, pp. 6–11, 82–86.

17. Rostasy, F.S., personal communication.

18. Meier, U. and Deuring, M., The application of fibre composites in bridge repair. *Strasse Verkehr* **77** (1991) 775.

8 Aramid-based prestressing tendons
A. GERRITSE

8.1 Introduction

8.1.1 General remarks

Structural concrete and cement-based materials have been extensively used in the building industry for about a century, because they are cheap, strong (in compression) and mainly durable. To improve some of the weak points, such as restricted tensile capacity, or corrosion of reinforcement, waves of developments are appearing with a certain regularity. These developments, sometimes with claimed 'fabulous' characteristics, open up very effective and imaginative improvements or extensions of the range of applications of cement-based materials. They should, however, be treated with a considerable amount of care or even suspicion, especially concerning their long-term behaviour. This will also be applicable to new structural materials presently being developed. A sensible equilibrium between care (suspicion) and imaginative use must be based on a reliable amount of knowledge of durable behaviour, which means interpretation and extrapolation of experimental data, as well as on some engineering judgement. It should be remembered that the effects of improving a weak point (e.g. corrosion) by introducing new materials can be reduced by the introduction of other, unknown, unexpected or less-wanted phenomena (e.g. insufficient stability in alkaline environments).

Concrete elements or concrete structures are commonly reinforced with steel reinforcement and/or prestressed with prestressing steel to accommodate tensile forces. Although steel has proved to be a high strength, reliable material, it could be advantageous for some applications, e.g. concrete elements or structures in highly aggressive environment, to use a combination of concrete with alternative means of reinforcement.

The alternatives to steel reinforcement can be divided into: (i) metallic solutions (epoxy-coated, galvanised or stainless steel); and (ii) non-metallic solutions, mainly based on fibres (asbestos, glass, polypropylene, aramids, etc.). This chapter deals with continuous fibres (fibrous tensile elements).

Recently, bars and tendons have been developed which consist of continuous high-strength, high-modulus fibres, usually—but not necessarily—embedded in a polymeric matrix. These fibres are designed for structural application in concrete [1] and comprise:

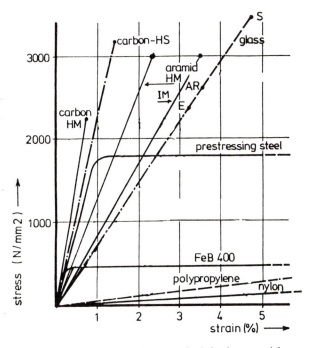

Figure 8.1 Stress–strain diagrams of reinforcing materials.

- glass fibres;
- aramid fibres;
- carbon fibres.

The strength capacities of such fibres are in the range of prestressing steel [2] (Figure 8.1). However, several other material characteristics differ considerably from those of the well-known steels. This is especially important for those less familiar phenomena or characteristics which would not be expected to be critical or decisive. The assumed common practice in the near future, for use of these materials by structural engineers, necessitates the need for reliable data specifically with respect to their long-term behaviour [3, 4]. This will necessitate investigation and evaluation of the behaviour as well as the mechanical, physical and technological properties in relation to long-term use, specifically in concrete structures. Design Criteria and Rules will have to be developed with emphasis on:

- behaviour under sustained stress and in aggressive (alkaline) environments;
- ductility of the structure (warning of failure).

The continuous fibre alternatives can appropriately be applied in the case of:

- very thin concrete elements;
- aggressive or corrosive environments (e.g. seawater, splash-zones, de-icing salts);
- need for non-electrical or non-magnetic conductive materials.

However, they must comply with the essential performance criteria expected from concrete structures.

Satisfactory application of tensile elements can only be achieved if basic requirements are fulfilled. These include:

- adequate physical and mechanical properties;
- durability in environments occurring in actual practice;
- ability to transfer force.

Concrete elements and structures are expected to last for many years (a 'lifetime') and to perform adequately during this period—say 50 to 100 years. This means *continuous fulfilment of the function for which the elements or structures are designed ('fitness for use')*.

An additional argument for defining methods to extrapolate long-term behaviour may be found in the less positive behaviour recently experienced with epoxy-coated steel bars [5]. Practical use of new materials must be based on reliable extrapolation.

The aim of this chapter is to discuss the recent developments in and applications of aramid fibres, with emphasis on Arapree, an aramid/epoxy-tendon based on Twaron* and jointly developed by AKZO and HBG. Arapree is produced using Twaron by SIREG SpA in Aracore, Italy.

Several variants of the modulus of elasticity of aramid are possible. The main experience to date, as described in the following sections, is based on the use of the HM (high modulus) type (see Section 8.2). Recently, a cheaper variant of Arapree based on Twaron IM (intermediate modulus, see Section 8.2) has become available and is expected to become the common standard.

8.1.2 *Material development*

Although the aim of this chapter is to indicate the potentials of and experience with aramid-based tensile elements, it is also appropriate to include a summary of the main arguments for the use of non-metallics, and the range of products available or under development.

The main properties generating interest in the use of advanced fibres as an alternative to steel in reinforced and prestressed concrete structures are:

- High strength—up to 3000 N/mm^2 (on net fibre cross-section);
- High modulus—similar to known steel equivalents, to get deformations of the same order;
- Non-corrosive—all fibres are suitable in carbonated environments. Carbon also exhibits a good resistance to both highly acid and highly alkaline environments. Glass fibres are less resistant to a high degree of alkalinity. The chemical resistance of Arapree will be dealt with in later sections;
- Resistant to aggressive environments such as chlorides;
- Insensitivity to electro-magnetic currents (this does not fully apply for carbon fibres);
- Excellent fatigue behaviour (except glass);
- Influence of temperature on relaxation behaviour nearly negligible— in contrast to prestressing steel, where relaxation figures strongly increase with moderate temperature rise.

The practical use of these new fibres as tensile elements in concrete is restricted by:

- The lack of experience with and hesitation to use 'non-proven' materials;
- Relatively low E-modulus (except for carbon fibres). This low ratio between the stiffness and the strength restricts the applicability of the new fibres as passive reinforcement in concrete;
- Absence of plastic deformation. An almost linear relation between the stress and the strain up to failure is observed (Figure 8.1);
- State-of-the-art with respect to the development of adequate anchorages capable of prestressing the tendons;
- Price level.

Despite these restrictions, it is worthwhile investigating and exploiting the advantageous properties, particularly in view of the improvement in the durability of concrete in 'exposed' conditions [6].

An indicative list of some products being developed for commercial use is given in Table 8.1. The range of types under development is expanding very rapidly. In all except one case continuous fibres, mainly parallel lay, are used; the exception is the braided rods of Mitsui (Fibra) [7].

Two methods of bundling the fibres to form practical units are utilised.

- bonding with resin into bars, rods, strips, etc. (parallel lay or braided);
- bare fibres encased in a sheath (parallel lay).

There are several reasons for embedding the fibres in a resin, including:

- handling of the elements;
- optimal use of fibre strength;

Table 8.1 High-strength fibrous tensile elements: indicative listing

Tensile element	Type of fibre	Fibre trade name	Type of composite	Producer	Reference
Arapree	Aramid	Twaron-HM, IM	Epoxy-resin bonded bar	AKZO (NL); Sireg S.p.A. (Italy)	[8]
Fibra-rod	Aramid	Kevlar 49	Braided rod in epoxy resin	Mitsui (Japan)	[7]
				Shinko Wire Co.	
AFRP	Aramid	Technora	Vinylester bonded bar	Teijin (Japan)	[9]
Parafil	Aramid	Kevlar 49	Fibres in sheath	ICI (UK)	
Polystal	Glass	E-glass	Polyester-resin bonded bars and cables	Linear Composites	[10]
Bri-ten[a]	Aramid	Kevlar	Resin-bonded cable	Strabag (Germany)	[11]
	Carbon	Carbon HS		Bridon (UK)	[12]
Nefmac	Glass		Resin-bonded meshes	Nefcom (Japan)	[13]
	Aramid			Shimizu	
	Carbon				
CFC	Carbon		Epoxy-resin bonded bars and cables	Tokyo Ropes (Japan)	[14]
Lead Line	Carbon		Epoxy-resin bonded bars	Mitsubishi Kasei (Japan)	[15]
Carbon stress	Carbon		Resin-bonded bars	Nederslandse Draad Industrie (NL)	[16]
GFCC	Glass	S2-glass	Resin-bonded bars	Owens Corning (USA)	[17]
Kodiak	Glass		Resin-bonded bars		[18]
Jitec	Glass		Resin-bonded bars		[18]
	Carbon				

[a] Commercial development ceased.

- transmission of internal shear stresses in the anchorage zone and in the vicinity of an accidental filament rupture;
- improvement of chemical resistance;
- ultraviolet protection. If used in concrete this aspect does not play any role. Only in case of external prestressing might it be significant.

Impregnation of the bundles means that a break in an individual fibre will lead only to a marginal local increase of the stress level in the neighbouring fibres. Local shear stresses in the resin will compensate for this and cause a nearly uniform stress level in the cross-section. Thus, by impregnating the fibres optimal use of the fibre strength can be achieved and the true tensile strength can be activated. Strength values (short-term and long-term) of non-bonded (bare) bundles with the same net fibre cross-section are about two-thirds of those of identical ones in resin-bonded elements.

It should be recognised that 'resin/fibre-bars' are very vulnerable to the common tooth-type steel grips. Temporary and permanent gripping to induce (pre)stressing forces has proven to be difficult. Specifically, anchorage devices for larger (cable-like) units as used in post-tensioning are problem items in development. The wedge-type anchorages developed by HBG [6] have proven to be a suitable tool for smaller units.

Arapree development is presently aiming at pretensioning.

The choice of resin will influence the speed and hence the cost of production of the bars or tensile elements. However, durability aspects are also involved, as the rate of ingress of ions aggressive to the fibres (e.g. hydroxyl ions) is strongly involved [19].

8.2 Aramid-based tensile elements

8.2.1 Aramid

8.2.1.1 Structure. Aramid is an organic, man-made fibre with a high degree of crystallinity. The molecular chains are aligned and stiffened by means of aromatic rings linked by hydrogen bridges. This combination explains the very high strength.

Aramid is a generic term used to describe fibres from wholly aromatic polyamides. These are aramid polymers in which at least 85% of the aramid linkages are attached directly to two benzene rings in the polymer chain. Figure 8.2 shows the structural formula of an aramid, which has been produced from *para*-phenylene diamine and terephthaloyldichloride by polycondensation.

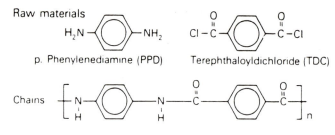

Figure 8.2 Structure of aramid fibres.

8.2.1.2 *Producers and products.* Aramid fibres are produced by:

- Dupont de Nemours in several plants (brandname Kevlar);
- Aramide Maatschappij vof (AKZO + NOM) in the Netherlands (brandname Twaron);
- Teijin in Japan (brand name Technora).

The first two offer aramid in several grades of elastic moduli (ranging from approximately 70 to 200 kN/mm^2). The filaments normally have a diameter of approximately 12 μm, which means a cross-section of 0.11×10^{-3} mm^2 (100 000 filaments $= 11$ mm^2) [6].

Technora is only available with an elastic modulus of approximately 70 kN/mm^2 and has a marginally different chemical composition from the former two. The same general mechanical characteristics apply.

8.2.1.3 *Characteristics.* The average tensile strength of aramid fibres is approximately 3000 N/mm^2 or even higher, measured as filaments. This means approximately 33 kN per 'unit' of 100 000 filaments, which can only be reached by fibres embedded in resin. With unbonded fibre bundles this is impossible. For practical purposes a characteristic strength of 2800 N/mm^2 may be reliably assumed, though with a large scatter in the data. This leads to a characteristic strength of 30 kN per 'unit' containing 100 000 filaments. For relevant material properties Section 8.4 should be consulted.

There are three types of Twaron aramid: (i) normal (NM); (ii) intermediate (IM); and (iii) high modulus (HM). The moduli ranges can be defined as:

- NM approximately 70–80 kN/mm^2;
- IM approximately 90–100 kN/mm^2;
- HM approximately 120–130 kN/mm^2.

8.2.1.4 *Some historical notes.* One of the first publications on non-metallic tensile elements (the Rubinski's) dates back to 1954. This is

related to glass fibres [20]. Continuous reinforcement using aramids was first investigated by Cleary. Later, Hughes tested GRC troughs to which bare continuous aramid strands were added. These tests were not as effective as assumed [21] and the attempts were not continued. Den Uijl and Reinhardt conducted preliminary small-scale investigations on impregnated aramid yarns [22]. Burgoyne of the Imperial College of London investigated Parafil ropes for use as unbonded tendons in concrete structures [23]. This is discussed in Chapter 5. Another notable pioneer is Dolan [24].

It should be mentioned that the fields of application of Twaron in civil engineering are not restricted to use in concrete. A product such as Twaron-based cables is also worthwhile considering for use in ground anchors in aggressive environments.

8.2.2 Arapree

8.2.2.1 Structure. Arapree stands for **aramid prestressing element** and is a joint development by AKZO and HBG (Hollandsche Beton Groep). It consists of Twaron filaments embedded in epoxy resin. This embedment facilitates handling, improves alkali resistance, and activates the true tensile strength of the material. The cross-section is made up of approximately 50% aramid and 50% epoxy.

Tendons have been developed which consist of parallel filaments of aramid and a simultaneous impregnation by epoxy resin. The bundles of Twaron in an Arapree tendon are non-twisted. The tendon is formed by passing the fibres through eyelets and combs and subsequently impregnating them with an epoxy resin.

Arapree is a composite of many thousand of filaments of Twaron, and an epoxy resin. Presently available types of Arapree are summarised in Table 8.2. There are two types of Arapree elements, one with a rectangular cross-section and one with a circular cross-section, with up to

Table 8.2 Available sizes of Arapree

Type[a]	Shape of cross-section	Size (mm)	Aramid fibre cross-section (mm^2)	Characteristic strength (kN)
f 100 000	Rectangular	20 × 1.5	11.2	30
f 200 000	Rectangular	20 × 3.0	22.2	60
f 400 000	Rectangular	20 × 6.0	44.4	120
f 100 000	Circular	ϕ5.7	11.1	30
f 200 000	Circular	ϕ7.9	22.2	60

[a] Equal to the number of filaments.

400 000 filaments of aramid. For example, the ultimate tensile strength of 200 000 filaments is about 67 kN. To avoid misunderstanding which may arise due to the relatively large shape-depedent or production-method-dependent variations of resin content, the mechanical properties are based on the (aramid) fibre cross-section. The available length of the prestressing elements is almost unlimited due to delivery on coils. The round bars can be smooth (nearly no bond) or sanded with a good bond performance. The rectangular bars (strips) are produced with a slightly profiled surface.

Due to the production method chosen for the rectangular and profiled strip, bonding between the prestressing elements and concrete appears to be very effective. The anchorage length of rectangular Arapree strands (strips) with profiled surface f 100 000 in a C45 grade concrete was only 50 mm for full prestressing force. Actual tests proved that even 50 mm was sufficient up to the breaking load of strands (approximately 35 kN).

For specific clients a range of circular wires with less than 100 000 filaments (diameter less than 5.7 mm) is produced, down to a diameter of approximately 1 mm. Such variants are collectively named Aracore. Since these sizes are less relevant to concrete structures they will not be discussed further here.

Arapree (and Aracore) is currently produced (from Twaron fibres) by Sireg S.p.A. in Italy. Sireg is also applying Arapree bars in fields outside concrete structures, e.g. ground anchorages, strengthening of old brick work, etc.

In general the influence of the resin on the characteristics of the bars is nearly negligible (except for durability). However, special attention should be given to thermal expansion values of the resin, with emphasis on the radial effects (see Section 8.4.2.2). Arapree circular bars can be produced with a foamy coating to overcome this drawback.

8.2.3 Fibra-rod

Fibra-rod is a development of the Mitsui Construction Company, intended to reinforce concrete. In cooperation with Shinko Wire Co., Mitsui also aim to use fibra-rod to prestress concrete. Fibra-rod is produced by braiding continuous aramid fibre (Kevlar) and impregnating it with epoxy resin. The strength and elastic modulus data are in the same range as the Twaron HM based Arapree tendons (2800 N/mm^2 and 130 kN/mm^2, respectively, see Section 8.4.2). To provide the bond, the surface is commonly sanded.

A range of (circular) cross-sections is listed [7], from 3 mm diameter, type K8 with a nominal strength of 8 kN, to 16 mm diameter, type K256 with a nominal strength of 256 kN.

8.3 Structural criteria

From the practical point of view of application, as well as from the structural point of view when designing with adequate safety for an adequate life span, the following items must be considered.

(1) *Physical properties*
  - Density
  - Poisson's ratio
  - Coefficient of thermal expansion (axial *and* radial)
  - Conductivity (electrical and magnetic)
  - Water uptake

(2) *Mechanical and technological properties*
  - Stress–strain diagram
    - shape;
    - strength;
    - strain at failure (fracture mode);
    - stiffness (elastic modulus).
  - Long-term behaviour under sustained load
    - stress rupture (static fatigue)/residual strength;
    - creep (stress decay);
    - relaxation.
  - Fatigue behaviour

(3) *Durability, behaviour in different environments*
  - Cementitious matrix/residual strength
    - alkaline environment (pH > 12);
    - carbonated concrete (pH < 10).
  - Presence of chlorides ($CaCl_2$, seawater, de-icing salts)
  - Other aggressive environments (SO_3, pure H_2O, etc.)
  - Temperature effects (very high, very low)
  - Stress corrosion
  - Ultraviolet resistance

(4) *Force transfer*
  - By bond (direct contact with cementitious matrix)
    - fatigue of bond
  - By anchorage (full force to anchorage device)
    - temporary (mainly pretensioning)
    - remaining during service life (mainly post-tensioning)
  - Secondary reinforcement

(5) *General*
  - Cost effectiveness
  - Construction aspects

8.4   Properties and phenomena

8.4.1   General

8.4.1.1   Aramid-based tendons, Arapree.   The items listed in Section 8.3 show several phenomena relatively unfamiliar to engineers using steel to reinforce or prestress concrete. Designing with (artificial) fibrous elements must include as many reliable data on these phenomena as possible. In the following sections present knowledge and relevant considerations will be discussed. Some values are the result of the combined effect of fibres and resin (e.g. Poisson's ratio and thermal expansion). Others are specifically related to the net cross-section of the fibres only, because variations in fibre/resin content, originating from production requirements, will lead to different figures (e.g. strength values). In the author's opinion this is a realistic approach; however, many workers use combination values given by the producers, which results in a range of different values.

8.4.1.2   Resins.   For practical reasons, such as ease of handling, good adherence to a variety of fibres, and stability and resistance to many chemicals, epoxy resin was selected as the bonding matrix to produce Arapree tensile elements. These, particularly when in direct contact with a cementitious matrix, are unrivalled in meeting the technical requirements for structural applications. Several types of epoxy resin have been investigated for their workability and effects on durability (see Section 8.4.5).

Other fibrous tensile elements are produced with polyester resin or vinylester. For this category, additional protection in aggressive (alkaline) environments is often unsatisfactory as ingress of ions is insufficiently retarded. To the author's knowledge, comparative testing, using the same fibres, between vinylester and epoxy-resin embedded fibres has not been carried out.

The elastic modulus of the resins used is so low compared to that of the fibres, that the part of the load carried by the resin may virtually be neglected. However, although loads are carried by the fibres, the effects on durability (Section 8.4.5) and radial thermal expansion (Section 8.4.2.2) must be considered.

8.4.2   Physical properties

Physical properties relevant to design with Arapree are indicated in Table 8.3. The properties for comparable tendons of other materials must be defined.

Table 8.3 Physical properties relevant to design with Arapree

Property	Aramid Twaron	Arapree (Twaron + epoxy)	Units
Density	1450	1250	kg/m^3
Coefficient of thermal expansion			
Axial	-5	-2	$10^{-6}/°C$
Radial	Approx. $+50$	Approx. $+50$	$10^{-6}/°C$
Electrical conductivity			
Dry	–	7×10^{-15}	Ohm $\times$ cm
Wet	–	7×10^{-7}	Ohm $\times$ cm
Poisson's ratio	0.38	0.38	

8.4.2.1 Poisson's ratio. This value is much larger than that for steel reinforcement. This may, for example, indicate larger splitting forces at release of prestress using these elements.

8.4.2.2 Coefficient of thermal expansion

Axial. The coefficient of longitudinal (axial) thermal expansion of aramid is slightly negative. Combined with the positive value of the resin $(50$ to $80 \times 10^{-6}/°C)$, the value approaches zero, due to distribution by the relevant elastic moduli. This leaves a difference from the surrounding concrete of approximately $12 \times 10^{-6}/°C$.

Under ambient conditions concrete structures can accommodate this. However, a marginal increase in prestressing force—dependent on the temperature rise—should be given consideration at higher temperatures.

Radial. The radial coefficient of thermal expansion causes concern [19, 25]. This is one of the unexpected—and unwanted—characteristics of the new materials (see Section 8.1.1).

The coefficient of radial thermal expansion of aramid is positive and large. Combination with the positive value of the resin results—dependent on fibre content—in values up to 4 or 5 times that of the surrounding concrete. This can lead to considerable splitting stresses causing longitudinal cracks, even due to a moderate temperature rise. The solution, for concrete covers less than 20 mm, is a (patented) method of covering the bars with a thin foamy coating. This releases the internal stresses around the tensile elements caused by temperature differences.

It should be noted that this phenomenon is not restricted to aramid or Arapree. It is basically a problem caused by the physical property of the resin and will also be experienced—to different levels—with carbon/resin and glass/resin combinations. Presumably it might even be present if using epoxy-coated steel tensile elements.

Table 8.4 Short-term mechanical properties of Arapree

Property	Average values		Dimension
	HM	IM	
Axial tensile strength[a]	3000[b]	3000[b]	N/mm^2
Modulus of elasticity[a]	125–130[c]	90–100[c]	kN/mm^2
Failure strain	2.3	3.5	%
Transverse compressive strength	Approx. 150	Approx. 150	N/mm^2
Interlaminar shear strength	Approx. 45	Approx. 45	N/mm^2

[a] Values related to the effective fibre cross-section.
[b] Characteristic value 2800 N/mm^2.
[c] Modulus based on measurements in the range from 0.35% to 1.35% of strain (Figure 8.3).

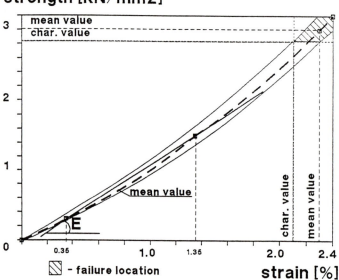

Figure 8.3 Stress–strain relations of (HM-based) Arapree.

8.4.3   Mechanical properties

8.4.3.1   Stress–strain behaviour.   Figure 8.3 and Table 8.4 show the characteristic values of stress–strain relations of (HM-based) Arapree. Curves for IM-based bars are comparable, with the same strength, but higher strain. The short-term values are derived from a series of tests of the material as produced. However, as can be seen from the following sections, serious reductions in the short-term strength values must be taken into account due to phenomena such as durability under sustained stress in the presence or absence of aggressive (e.g. alkaline) environments [26]. The real stress–strain curve has a very slight upward curvature. The data given for the E-modulus are the angles measured between strains of 0.35% and 1.35%. Characteristic values may be assumed to be those—as common in civil engineering—limiting maximum 5% lower measurements.

Strength. Strength data of all aramid products are basically of the same level. With the optimal stress transfer of the resin matrix used values of over $3300\,N/mm^2$ are reached with Arapree.

Strain at failure. Aramid, and consequently Arapree, has, like the whole range of fibre-based tensile elements, no yield range. It is brittle, and this must be carefully considered by designers using the ductility approach commonly included in the design of concrete structures. The expected warning behaviour (deflections etc.) can be different or even absent, especially for those artificial fibres with even lower strain capacity, e.g. carbon-fibre based tensile elements [27].

8.4.4 Technological properties

8.4.4.1 Creep and relaxation: influence of time.

General. Relaxation and creep are interrelated material properties. Both describe a relation between stress, strain and time. Creep describes a change in strain as a function of time at constant stress. Relaxation describes the change in stress as a function of time at a constant length. The former is usually expressed in direct strain figures indicating the increase, while the latter is expressed as percentages of the initial stress.

The reaction of the structure of the materials, on being stretched, is specific. No general law gives a fixed relation between creep and relaxation. A significant difference between prestressing steel and polymeric materials becomes apparent from tests at different stress levels [28].

Creep. Polymeric materials show an approximate *linear relation* between the applied stress and the creep strain at all times [28].

Prolonged measurements indicate a more or less straight line in the log/log plot as extrapolated in Figure 8.4. The values given in the figure are determined at an initial stress level of about 50% of the characteristic strength. After $10^6\,h$ the expected creep strain is approximately 0.2% in air. The creep coefficient is 0.2 at $10^6\,h$.

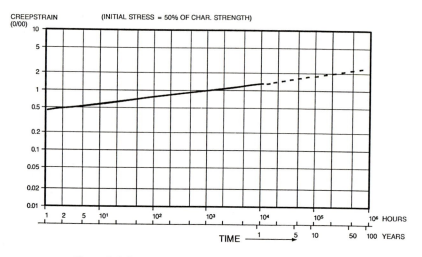

Figure 8.4 Creep behaviour of Arapree HM in air at 20°C.

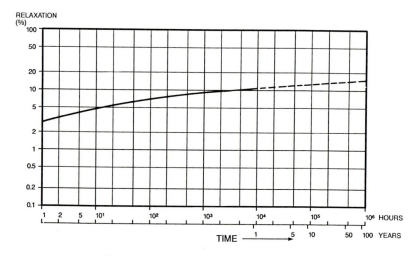

Figure 8.5 Relaxation of Arapree HM in air.

Relaxation. The relaxation of polymeric materials appears to be more or less *independent of the applied stress level* [28]. Figure 8.5 shows the losses of stress due to relaxation—independent of stress level—for Arapree in air.

Comparison between creep and relaxation of steel and aramid. Three possible relationships between creep and relaxation can be assumed.

From steel it is known that the choice of a higher initial stress level gives an increase in creep strain which is greater than proportionality (Figure 8.6(a)). Relaxation data show comparable results.

Investigations with aramids, however, show that these have different behaviours. With an increase in stress—as a percentage of ultimate strength—creep strain increases proportionally or even less than proportionally, from ε_{cr1} to ε_{cr2} (Figure 8.6(b), (c)). It can be concluded that—contrary to prestressing steel—an increase in initial stress produces *constant* or even *decreasing* relaxation percentages for aramids over the same time interval [28]. For practical design a *constant* relaxation percentage may be assumed to be valid (Figure 8.7). Consequently, relaxation at t_x can be obtained from any creep test in the practical stress range. The apparent long-term modulus E_{tx} gives the relation (Figure 8.6(b)).

Influence of moist environments. The influence of dry or wet atmospheres is evident from Figure 8.8.

Influence of temperature. Creep and relaxation have proved to be almost independent of temperature, which again might be surprising for polymeric material. This can be an important feature since relaxation of prestressing steel increases very strongly at a moderate temperature rise [28].

Summary. There is an apparent difference between the creep and relaxation behaviour of prestressing steels and aramid-based tendons. The losses of the tendon (10^6 h) under non-ideal (wet) conditions are of the same order of magnitude as expected from the prestressing steels used up to 15 years ago (relaxation type I). However, compared to the current low relaxation steels (type II), the values are relatively high.

The level of creep or relaxation for Arapree is virtually independent of the level of prestress and is almost insensitive to temperature differences. Furthermore, due to the lower elastic modulus, the effects of shortening of the concrete (creep and shrinkage) are compensating. Thus, the final losses to be taken into account in the concrete design may still be of the same order.

It should be noted that most investigations on creep/relaxation have been carried out on Arapree based on Twaron HM [29].

8.4.4.2 Stress rupture: effect of sustained load. For most materials subjected to sustained relatively high stress (as is common in prestressed concrete), the failure strength decreases with time (long-term strength). This is commonly not critical for prestressing steel, but the rate of the decrease, and the available residual strength just before failure, is a priority criterion for polymeric (organic) materials. When this phenom-

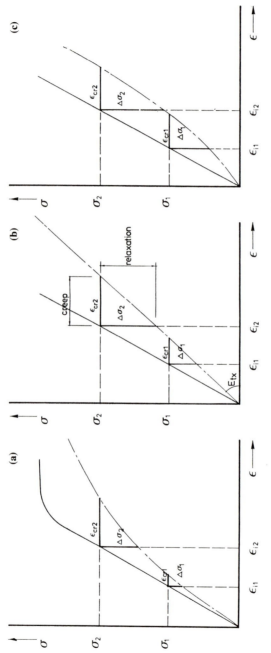

Figure 8.6 Relationships between creep and relaxation of: (a) steel; (b) and (c) aramid. Solid lines indicate initial response, broken lines indicate final response.

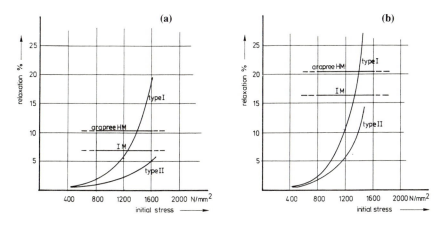

Figure 8.7 Relationship between stress and relaxation of aramids: (a) measured at 10^3 h; (b) anticipated after 10^6 h.

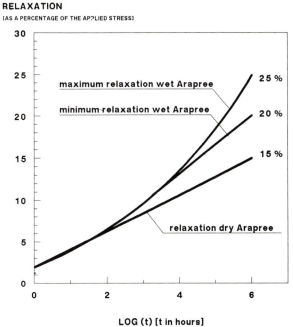

Figure 8.8 Relaxation of Arapree—effect of moisture.

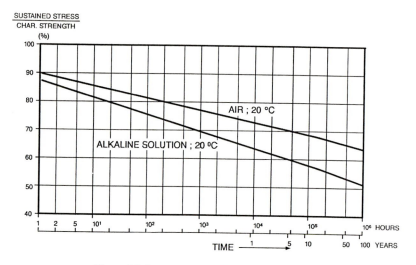

Figure 8.9 Stress rupture behaviour of Arapree.

enon occurs in air it is called stress rupture (Figure 8.9). Stress corrosion is the same phenomenon occurring in a particular chemical environment (e.g. an alkaline environment).

For a certain sustained load level in the fibres stress rupture behaviour predicts the relation between this level and the time it takes for a tendon to fail under the corresponding *constant* stress. Stress rupture may also be called static fatigue or long-endurance strength, and this can be compared with the compressive strength of concrete under sustained loading.

As can be seen from Figure 8.9 the time to failure—under a certain load—depends on the type of environment. A load causing stresses of 60% of the short-term strength generally has a survival time of over 10^5 hours in an alkaline environment. However, the figure also shows that a sustained load of 50% of the short-term breaking load will not lead to failure.

Measurements of stress rupture in a saturated alkaline solution may be taken as a very severe representation of hardened concrete. Measurements of stress rupture or stress corrosion always have a relatively large dispersion on the time scale (Figure 8.10). Consequently, characteristic limits must be defined.

It should be noted that the short-term ultimate strength—needed at a sudden overload—is almost independent of this phenomenon. Thus, if an Arapree tensile element is stressed to, say 60%, sustained for 10^4 h and then suddenly short-term *additionally* loaded it will show approximately the original short-term strength.

It should be remembered that the data given here are extrapolated

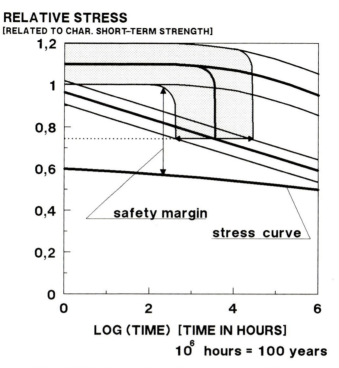

RELATIVE STRESS
[RELATED TO CHAR. SHORT–TERM STRENGTH]

LOG (TIME) [TIME IN HOURS]
10^6 hours = 100 years

Figure 8.10 Safety margin in alkaline solution at 20°C.

from research, sometimes using an accelerated test. The acceleration is performed by increasing the temperature (see Section 8.4.5).

8.4.4.3 Fatigue behaviour. The fatigue behaviour of Arapree has proved excellent, which is in accordance with extensive international literature. This is illustrated by Figure 8.11, which compares results of fatigue tests on Arapree with the required fatigue strength of prestressing steel. Additional measurements by den Uijl [26] confirm this.

8.4.4.4 Bond. Arapree behaves very satisfactorily when in direct contact with the concrete matrix, both in carbonated and in alkaline environments (see Section 8.4.5). This makes it very suitable for use as reinforcement or, in the case of pretensioning, as prestressing elements. Concrete elements or structures without metallic parts are thus becoming possible.

The forces acting on the cross-section of prestressing tendons are relatively very high. Consequently, for the transfer of forces from the centre of the rod to the surface (where it is held) a large surface area in relation to the cross-section is to be recommended. Originally it was

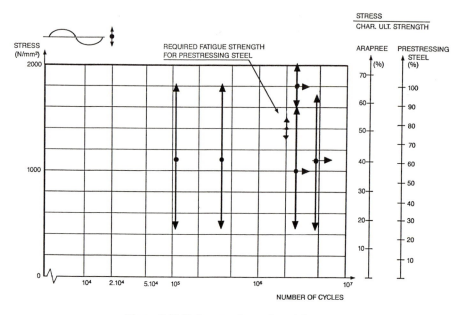

Figure 8.11 Fatigue test in tension of Arapree.

decided to develop a strip-like or rectangular tendon, mainly to improve the transfer of forces. This shape, which has an additional pattern of small knobs, proved to be very effective in bonding/direct contact with the cement matrix. For practical reasons circular shapes have also been produced, lightly sanded to allow transfer of forces by bond. Smooth elements provide nearly no force transfer.

With the pull-out test the bond slip curve for short embedment length can be determined, which serves as a basis for the calculation of transmission length [26]. Bond strength and the bond slip relation depend primarily on the following properties of the tensile element:

- roughness and treatment of surface (smooth, profiled, sanded);
- ratio of surface area to cross-section;
- thickness, shear strength and shear modulus of the near-surface layer of the matrix resin, etc.

The bond behaviour of Arapree in concrete has been tested on several different surfaces. For flat profiled shapes and sanded circulars transfer lengths of 50–100 and 150–300 mm, respectively, proved to be sufficient with failure in the element.

Bond behaviour is also important for non-metallic elements used as the reinforcement of concrete members which are liable to cracking under service load [30]. In addition to other parameters, the bond slip curve and the Young's modulus of the tensile element influence the crack width (which in itself is irrelevant because Arapree does not corrode). The test results of fatigue on the tendons [26] also indicate outstanding fatigue behaviour. These tests were carried out with the ends of test bars embedded in concrete elements.

8.4.5 Durability in the environment

8.4.5.1 Alkaline environments

General. It is essential for the tensile elements to be used in prestressed/pretensioned concrete to retain sufficient strength capacity, during the normal lifetime of the concrete structure, in any of the environments which might be encountered in practice. In the case of direct contact, embedment in a cementitious matrix will result in the first instance in alkaline exposure (pH > 12) as well as later on, possibly, carbonated concrete (pH < 10). The major argument for using non-steel, tensile elements is that steel corrodes in carbonated concrete; this will be seriously enhanced if chlorides are present (e.g. coastal areas, de-icing salts, etc.).

Alternative materials based on glass, polymeric or carbon-fibres, generally behave satisfactorily in carbonated concrete; however, some, particularly glass, will deteriorate in an alkaline environment. Durability (or residual strength) of any fibre in alkaline environments will therefore define its practical applicability in concrete elements, if in direct contact with the cementitious matrix [31]. Figure 8.12 shows a schematic indication of degradation in solutions with different pH values.

Arapree tendons also exhibit excellent resistance to chlorides and many other environments aggressive to steel. Specifically, the insensitivity to chlorides, such as de-icing salts, offers opportunities to overcome a range of existing deterioration problems in concrete structures. Even $CaCl_2$ can be added to the mix without detrimental effect to the 'reinforcement'. In such circumstances the behaviour of reinforcing or prestressing steel generally leads to additional requirements to improve and ensure the durability. If aramid is applied there will be no need for these measures.

Due to chemical attack the strength of Arapree may fall back to a level of about 85% of the initial strength (see Figure 8.10) after approximately 100 years.

Accelerated approach. It is obvious that the survival of alternative reinforcements in alkaline environments is an essential requirement for practical application in direct contact with concrete. The degradation of some fibres in alkaline environments is in principle governed by the

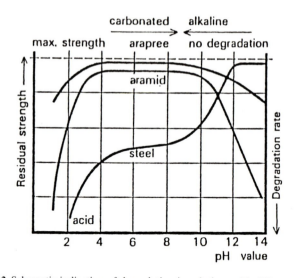

Figure 8.12 Schematic indication of degradation in solutions with different pH values.

attack from OH^- ions arising from the hydration process and reaching the fibres. This is a relatively slow process and the effects will only be observed after a number of years. To avoid such problems it is essential that accelerated methods are developed or defined to extrapolate the long-term behaviour. Recent evidence [3] serves to emphasise this requirement.

The accelerated approach is not new. The investigation and evaluation of 'accelerated ageing' has shown that the Arrhenius approach to this process is very sensible. The Arrhenius approach accelerates the process, by increasing the temperature (increasing the movement of the ions) and, provided no other disturbing processes occur (in the range from 0–100°C), it is a valid investigation.

Test method. The Arrhenius approach can be worked out mathematically by estimating the activation energy, which in this case indicates the temperature dependence of the processes of diffusion and attack on the fibres by OH^- ions. It can however also be carried out in a relatively simple graphical way, by inserting measured values in a graph such as Figure 8.13. At each temperature the process develops at a certain velocity. By measuring the residual strength values of samples at regular intervals, the rate of decrease in strength can be determined at each temperature. This will prove that, if a range of measurements—at regular time intervals—is carried out at several (minimum three) temperatures contour lines can be drawn connecting the same residual strength levels

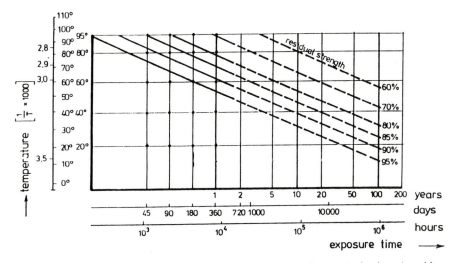

Figure 8.13 Extrapolation of residual strength plotted according to Arrhenius, Aramid epoxy tendon in saturated Ca(OH)$_2$.

(Figure 8.13). Using this approach the temperature indications on the vertical axis can be plotted on a linear scale with the factor $1/T$, where T is the temperature in degrees Kelvin.

The contour lines found ought to be straight to signify that no different—chemical—processes have interfered. Sloping contour lines indicate which residual strength can be expected at each chosen temperature level. Thus, if a significant part of these contour lines—which are basically parallel—can be established, an extrapolation of the rate of degradation to be expected (e.g. at room temperature in the given environment) can be determined from measurements at higher temperatures. If measurements over a sufficiently long period at higher temperatures are available the acceleration factor can be defined.

The alkaline resistance of the tensile elements (pure fibres, impregnated fibres of bar-shaped elements) can be evaluated by immersing samples in baths at different temperatures and with a certain—but maintained—alkalinity (e.g. pH value > 13). The contour lines shown in Figure 8.13 are determined on Arapree—a tendon based on aramid fibres in epoxy resin immersed in a saturated Ca(OH)$_2$ solution.

Several kinds of alkaline solutions can be assumed to be representative, from cement slurry to sodium hydroxide. The latter is largely overrated but, if a relation to reality can be investigated, it can give a very quick result. Some relation to the alkalinity of the pore water in the concrete has to be established.

The contour lines will shift to the left if less-resistant (e.g. glass) or less-protected (e.g. pure) fibres are chosen. The slopes can differ slightly depending on the particular resin chosen. However, from Figure 8.13 it can be concluded that an acceleration factor of more than 50 can be applied by increasing the temperature from room temperature to 60°C, and a factor of more than 200 by increasing the temperature to 80°C.

Discussion. Residual strength investigations comparable to the one given in Figure 8.13 can be made for different tensile elements and for different circumstances. Rather than estimating the capacity of the elements in a solution, it might be more appropriate to insert them in concrete products, which are then prestressed prior to immersion in heated water. This will give the most accurate information. In the author's experience, Arapree tendons in prestressed concrete elements show no sign of any strength decrease after 3 years exposure at 60°C.

Clearly the movement of ions in solution is much more active than in the concrete matrix. Results with glass tendons reported by Sen *et al.* [3] are much less positive.

From Figures 8.10 and 8.13 it is obvious that the tested aramid-based tendons retain more than 80% of their original capacity after 100 years at room temperatures. Since the slope of the contour lines will prove to be comparable in different cases it could be argued from the above that standardising a required measurement by investigating a single point (e.g. 180 days in alkaline solution at 80°C) can already indicate an extrapolated residual strength after 100 years real-time. This should become a standard.

8.5   Concrete structures

8.5.1   *Bonding behaviour/ductility*

When designing concrete structures the yield of the steel before failure offers several advantages. It ensures that there is a warning (e.g. visual deflection) if structures are likely to fail and it provides some rotational capacity, used for redistribution of moments in the case of serious overloading. Arapree has no yield behaviour (Figure 8.1). This however, does not imply that concrete structures reinforced or prestressed with Arapree lack the warning characteristics. They do not behave in a brittle manner.

In a beam or floor, with Arapree tensile elements calculated for the same live load combination, deflection will indicate overloading *earlier* (at a lower load) than if the beam or floor contained steel. The element or structure will not fail at that load, however. If prestressing steel is used the element will fail soon after the load reaches yield. With Arapree, deflection will continue even at higher loads. Partial coefficients (taking account of uncertainties in model and material characteristics) must be defined.

Design with Arapree tendons can be carried out according to the same approach as usual in concrete design, taking the relevant characteristics into account.

The nearly straight stress–strain 'curve' of aramid, lacking a ductile branch, does not give rise to the problems anticipated; warning behaviour is adequate.

8.5.2 Methods of anchorage

As already discussed (Section 8.4.4.4) Arapree behaves very satisfactorily in direct contact with the concrete matrix. However, to introduce the initial prestressing force into the tendons an anchorage—mainly for temporary purposes during construction—must be developed.

The strip-like or rectangular tendons, which were developed mainly to improve the transfer of forces and which proved to be very effective in bonding in direct contact with the cement matrix, also enabled simple anchorage devices to be developed (Figure 8.14).

The usual wedge-type steel grips in anchorage devices are not very effective on the resin surface of tensile elements. Polyamide wedges have proved to be adequate even at full force. Meanwhile, circular tendons,

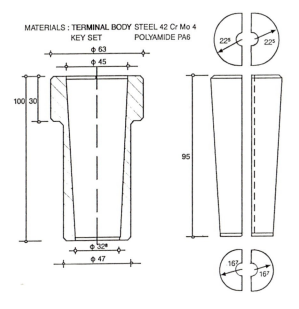

Figure 8.14 Simple anchorage devices.

as well as anchorage devices adapted to circular tendons, are being developed. Specific developments with devices to suit post-tensioning are not yet envisaged.

8.6   Applications and experience

8.6.1   Noise barrier posts

The first practical application of Arapree was a noise barrier along a busy motorway near Rotterdam. The ninety concrete posts of the barrier between which transparent sheets of PMMA are placed, are 4.5 m long and prestressed by rectangular Arapree elements. The post, with an effective cantilever of 3700 mm, is mainly loaded by wind forces and exposed to de-icing salt and spray water. The post is centrally prestressed with eight Arapree tendons. Under maximum wind load the concrete of the tensile zone is expected to crack. Since Arapree is not prone to chloride corrosion, the cracks are not dangerous.

8.6.2   Hollow-core floor slab

Another Arapree application is that of a hollow-core floor slab which is used in a project to demonstrate innovative developments in dwelling houses. Each floor slab has span of 6 m, with a standard width of 1.2 m and a depth of 0.26. The cross-section contained six Arapree prestressing elements of the rectangular type with an initial prestressing force of 400 kN per floor element.

8.6.3   Prestressed masonry

Prestressing of a masonry cavity wall helps to increase the bending resistance if only a little dead weight acts on top of the wall. In this case, the wall had a length of 40 m and a height of approximately 5 m. The rectangular tendons were placed every 330 mm and anchored in the foundation slab. After completion of the wall the tendons were stressed and anchored by bonding to the concrete cap beam. Due to the effective bond properties of the strip, the available transmission length of only 75 mm is sufficient. No corrosion protection measure is required in the cavity of the wall.

8.6.4   Fish ladders

Several hydroelectric power plants are currently operating in the Netherlands. To enable fish to pass these obstacles in the river, fish ladders are constructed. Normally these structures are made of tropical hardwood.

However, for environmental reasons this material is under discussion. Concrete planks prestressed with Arapree can be considered as a useful alternative. Approximately 2000 panels (1040 × 250 × 35 mm) have been produced, each prestressed with two Arapree bars. (The prestressing force in each panel is approximately 25 kN.)

8.7 Conclusions

In the emerging range of non-metallic tensile elements based on high-strength fibres, Arapree, the non-corrosive tensile element composed of parallel aramid filaments in an epoxy resin, may be considered to be a very promising alternative to reinforcing or prestressing steel for all applications where long-term durability is required. Aramid/epoxy tendons combine the qualities of very high strength with chemical resistance. The tendons are durable under practically every type of environmental attack which may occur in or around concrete structures. They can satisfactorily be used in prestressed or pretensioned concrete, as well as in direct contact with the cementitious matrix. An interesting feature is their insensitivity to large amounts of chlorides. From the experience available it can be deduced that the tendons will behave as required in bond and anchorage, fatigue behaviour, stress rupture, stress corrosion, etc. Moderately high and very low temperatures are not problematic. Concrete elements reinforced or prestressed with Arapree will not deteriorate as a result of the reinforcement. Cover is needed only for stress transfer and ultraviolet protection. Thus, very thin elements can be produced, while resistance to a range of specific effects on concrete structures such as impact or explosions will be improved.

Interesting application areas are:

- exposed concrete structures;
- presence of de-icing salts, splash zone, etc.;
- thin-walled elements with small concrete cover;
- replacement for tropical hardwood;
- exterior post-tensioning without corrosion protection;
- where fatigue is decisive;
- concrete with open (porous) structure;
- concrete containing salt (or $CaCl_2$);
- non-magnetic and non-conductive elements;
- radar installations;
- severely aggressive environments;
- permanent ground anchors.

References

1. Minosaku, K., Using FRP materials in prestressed concrete structures. *Concrete Int.* **August** (1992).
2. Gerritse, A., Prestressing with Arapree. *Symposium: New Materials for Prestressing and Reinforcement of Heavy Structures*, LCPC, Paris, 1988.
3. Sen, R., Isaa, M. and Mariscal, D., Feasibility of fiberglass pretensioned piles in a marine environment, College of Engineering, University of South Florida, August 1992.
4. Gerritse, A., Werner, J. and Egas, M., Developing design requirements for non-metallic tendons. *IABSE Symposium*, Brussels, 1990.
5. Sagues, A., Powers, R. and Zayed, A., Marine environment corrosion of epoxy coated reinforcing steel. *Corrosion of Reinforcement in Concrete*, Elsevier Applied Science, London, 1990, pp. 539–549.
6. Gerritse, A. and Schürhoff, H.J., Prestressing with aramid tendons, *FIP Congress*, New Delhi, 1986.
7. Commercial publication, Technical Research Institute Mitsui Construction Co. Ltd.
8. Reinhardt, H.W., Werner, J. and Gerritse, A., A new prestressing material going into practice, *FIP Congress*, Hamburg, 1990.
9. Commercial publication.
10. *Symposium on Engineering Applications of Parafil Ropes*, Imperial College, London, **January** (1988).
11. Weiser, M. and Preis, L., Kunstharzgebundene Glasfaserstübe, eine Korrosions-beständige Alternative für Spannstahl. *Fortschritte im Konstruktive Ingenieurbau*, 1984.
12. Walton, J.M. and Yeung, Y.C.T., Flexible Tension Members from Composite Materials. Publisher unknown.
13. Commercial publication.
14. Commercial publication.
15. Commercial publication.
16. Commercial publication.
17. Iyer, S.L. and Anigol, M., Testing and evaluating fiberglass, graphite and steel cables for pretensioned beams, *Symposium on Advanced Composite Materials in Civil Engineering*, ASCE, Las Vegas, 1991.
18. Clarke, J.L., Tests on slabs with non-ferrous reinforcement, FIP notes 1992/1 and 1992/2.
19. Gerritse, A., Durability criteria for non-metallic tendons in an alkaline environment, *Symposium Advanced Composite Materials in Bridges and Structures*, Sherbrooke, Canada, 1992.
20. Rubinsky, I.A., and Rubinski, A., An investigation into the use of fibre-glass for prestressed concrete. *Mag. Concrete Res.* **September** (1954).
21. Hughes, B.P., AGRC composites in thin structural sections. *Symposium: Advances in Cement Matrix Composites*, Boston, 1980.
22. den Uijl, J.A. and Reinhardt, H.W., Bond of aramid strands to cement mortar, *Symposium: Plastics in Building*, Liege, 1984.
23. Burgoyne, C.J. and Chambers, J.J., Prestressing with Parafil tendons. *Concrete* **October** (1985).
24. Dolan, C.W., Developments in non-metallic prestressing tendons. *PCI J.* **September/October** (1990).
25. Rojstaczer, S., Cohn, D. and Marom, G., Thermal expansion of Kevlar fibres and composites. *J. Mat. Sci. Lett.* **4** (1985) 1233–1236.
26. den Uijl, J.A., Mechanical properties of Arapree, Technical University Delft, Faculty of Civil Engineering (not publicly available).
27. de Sitter, W.R. and Gerritse, A., Elastoplastic design of composite structural members with brittle base materials, *Technologie und Anwendung der Baustoffe* (zum 60. Geburtstag von F.S. Rostasy), Ernst und Sohn, 1992.
28. Gerritse, A., Maatjes, E. and Schüthoff, H.J., Prestressed concrete structures with high strength fibres, relaxation, *IABSE Symposium*, Paris, 1987.

29. Rostasy, F.S., High-strength fiber composite tensile elements for structural concrete. State-of-the-art report of FIP commission on prestressing steels and systems (awaiting publication).
30. Faza, S.S., Behavior of fiber reinforced plastic rebar under bending and bond, *69th Annual Meeting, Transportation Research Board*, Washington, 1990.
31. Gerritse, A., Arapree a non-metallic prestressing tendon, *BCA Conference on Durable Reinforcement for Aggressive Environments*, Luton, 1990.

Index